MINISTRY OF
AGRICULTURE, FISHERIES AND FOOD

Irrigation

Bulletin 138

LONDON
HER MAJESTY'S STATIONERY OFFICE
1974

© *Crown copyright 1974*

First published 1947
Fourth edition 1974
Second impression 1976

Any reference to proprietary products in this bulletin should not be construed as an official endorsement of those products, nor is any criticism implied of products which are not mentioned.

ISBN 0 11 241438 9

Foreword

THE widespread adoption of irrigation in this country has been symptomatic of the developments in technology which our industry has undergone. This new edition of Bulletin 138 records the progress made.

The main sections of this edition have been written by Mr. L. P. Smith (Meteorological Office), Mr. E. J. Winter and Mr. T. Laflin (National Vegetable Research Station), Mr. G. P. Shipway (Agricultural Development and Advisory Service) and Mr. W. H. Hogg. Other contributions from A.D.A.S. have been by Mr. L. W. Wellings, Mr. J. Ingram, Mr. K. Mosdell, Dr. D. R. Smith and Mr. C. N. Prickett, together with contributions from Mr. R. S. Tayler (University of Reading), Professor J. P. Hudson (Long Ashton Research Station) and Dr. J. E. Goode (East Malling Research Station). We are most grateful to Mr. W. H. Hogg for overall editing.

Albert J. Davies
Chief Agricultural Officer
Agricultural Development and Advisory Service

Ministry of Agriculture, Fisheries and Food
January 1973

Contents

		Page
PART 1	GENERAL CONSIDERATIONS	1
	The Need of the Plant for Water	1
	The Determination of Soil Moisture Properties	5
	Meteorological Aspects	10
PART 2	PLANNING IRRIGATION	18
	Crop Response to Irrigation	18
	Water Supply—Sources	28
	Storage of Water	35
	Water Quality	36
PART 3	EQUIPMENT (OUTDOOR)	40
	Distribution Equipment—Types and Selection	40
	Pipes and Nozzles	45
	Pumps and Pumping	55
	Management of the System	58
	EQUIPMENT (INDOOR)	60
	Methods of Determining Requirement or Controlling Application	60
	Types of Equipment	64
	Systems of Irrigation	66
PART 4	IRRIGATION IN PRACTICE	73
	Irrigation for Crop Growth	74
	Farm Row Crops and Vegetables	77
	Nursery Stock and Flowers	84
	Fruit Crops	85
	Hops	88
	Grassland	90
	Glasshouse and Protected Crops	91
	Frost Protection	97
	Multi-Purpose Irrigation	102
	APPENDICES	103
I	Meteorological Data	103
II	Irrigation—Estimation of Frequency, Maximum and Average Needs	111
III	Planning Data for Specimen Plans of Irrigation	124
IV	Additional Irrigation Plans	126
V	Irrigation Costs	129
VI	Irrigation Technical Data	132
	Index	137

General Considerations

THE NEED OF THE PLANT FOR WATER

An acre of vegetation in full sunlight loses on average about 10 tons of water a day in summer; a square yard of crop uses about half a gallon of water a day. In irrigation terms, during high summer, crops in this country use an inch of water or more every 10 days.

Plants transpire these large quantities of water into the atmosphere during daylight and there is little or no transpiration at night. The water is extracted from the soil by plant roots, and the sun supplies most of the energy for turning liquid water into the vapour given out by the leaves. The transpiration stream has various important functions. The plant normally obtains its mineral nutrients from those dissolved in the soil water, which are taken up by the roots and transported in the transpiration stream to those parts of the plant where they are to be used. A very small proportion of the water taken up is combined chemically to form new plant material, by far the larger part being changed into vapour which passes into the atmosphere through pores (stomata) in the leaves. Each stoma is held open as long as its surrounding guard cells remain turgid; when these become flaccid because of water shortage in the plant they tend to close the pore and thus to some extent reduce water loss. Such reduction in the transpiration stream by stomatal closure leads to a decrease in plant growth.

Another essential function of the water taken up by the plant is to provide mechanical strength to non-woody tissues. The osmotic pressure of the solution inside the plant cells is such that they tend to take up so much water that they become turgid and stiff, thus imparting firmness to the tissue of which they form the component parts. If water loss from the leaves exceeds intake at the roots the cells become flaccid, the tissues are no longer self-supporting and the plant takes on the attitude which we recognize as wilting. The leaves and stems no longer maintain their proper posture for absorbing sunlight, the stomata close and thus hinder the ingress of carbon dioxide so that the process of photosynthesis and growth virtually ceases.

Plants normally obtain almost all of the water which is essential to their growth and well-being from the soil.

The contribution of so-called dewfall to the overall moisture status of the plant/soil system is very small. On the other hand, evaporation of dew during the day probably increases the humidity of the atmosphere around the plants, thus temporarily reducing the transpiration rate and easing any existing moisture stress.

The Soil Reservoir

The soil consists of particles of mineral matter (clay, silt and sand) and organic matter. Because of their irregular shapes these particles do not fit closely together and pore spaces are left between them. When rain or irrigation completely fills these spaces with water to the exclusion of all air the soil is said to be saturated, or waterlogged (Fig. 1). This is an unstable condition which is maintained only as long as drainage from the underside of the waterlogged layer is slower than the rate at which water continues to fall on its surface. When water ceases to fall on the surface, drainage tends to pull the water out of the pores and to replace it with air. Eventually, after a period which may range from a day to several weeks, according to soil type, the pull of gravity (which is constant) is balanced by the surface tension of the film of water around each individual soil particle. Drainage then virtually ceases and the soil is said to be at field capacity, that is, holding its maximum possible amount of water against free drainage.

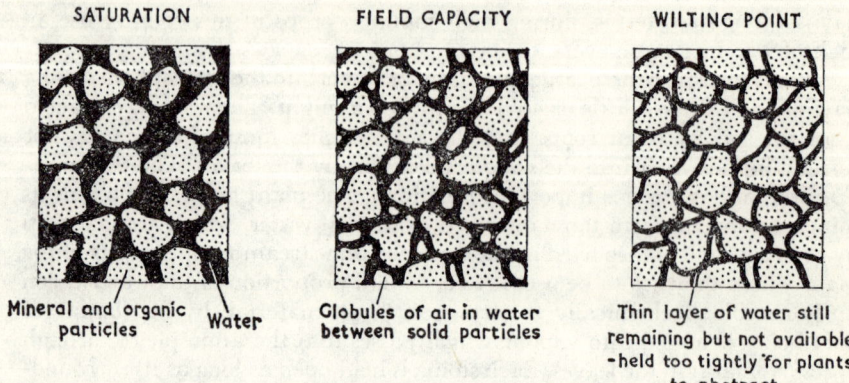

Fig. 1. Diagrammatic representation of terms used in describing moisture contents of soil

Field Capacity

The amount of water in the soil at field capacity varies for different soils. In fact very slow drainage continues from soils even after they have reached field capacity but at this point there is an abrupt fall in the rate of drainage. The best time to determine field capacity is in early spring after the thorough wetting of the soil which takes place during most winters.

No irrigation system should apply more water than is needed to restore the soil moisture status to field capacity. Any excess will be wasted in surface run-off or in drainage through the soil and there will be risk of soil waterlogging and of leaching of plant nutrients.

Drainage

Irrigation should never be carried out on poorly-drained soil; such a soil will become saturated in the normal way while the water is being applied but when watering ceases there will be no downward movement of water and consequent entry of air from the surface. The soil will remain waterlogged and incapable of supporting good plant growth.

Loss of Soil Water

Once it has reached field capacity the well-drained soil can lose water only by transpiration from plants growing in it or by evaporation from the wet soil surface exposed to the sun. So long as the soil surface is wet the rate of evaporation is almost the same as that from an open water surface. For practical purposes it can be assumed that not more than $\frac{3}{4}$ in. will be lost by evaporation from the soil surface before evaporation virtually ceases. Most of this loss is from the top few inches; capillary movement of water upwards takes place only very slowly.

Permanent Wilting Point

Unless a crop receives water from irrigation or rain, evaporation and transpiration reduce the soil moisture content until the plant roots can no longer take up water as fast as vapour is lost from the leaves. Internal water stress then arises with the consequential interference in important biological processes: the leaves wilt, transpiration is reduced and eventually ceases. When the plants remain wilted even in a humid environment the soil is said to be at permanent wilting point. In any given soil all normal plants wilt permanently at about the same soil moisture content.

A quite different form of wilting may occur in hot sun even when the soil is comparatively wet. This is the temporary wilting which may be seen in crops such as sugar beet, and when it occurs watering will not cause the plants to regain their turgidity while the intense evaporation continues.

Available Soil Water

Plants can take up water from soil at any moisture content between field capacity and permanent wilting point, and the quantity of water which a soil can supply between these two points is known as its available water capacity.

Available water capacity is a characteristic property of a soil although it can be changed to a small but useful extent by good husbandry such as suitable cultivation and the judicious use of organic manure. Table 1 shows the moisture properties of typical soils of this country.

Table 1
Soil moisture properties of typical soils

A LOW	Not more than $1\frac{1}{2}$ in. per ft depth
	Coarse sand
	Loamy coarse sand
	Coarse sandy loam
B MEDIUM	More than $1\frac{1}{2}$ in. but less than $2\frac{1}{2}$ in. per ft depth
	Loamy sand
	Sand
	Loamy fine sand
	Fine sand
	Sandy loam
	Fine sandy loams
	Loam
	Silty loam
	Silt loam
	Clay loam
	Sandy clay loam
	Silty clay loam
	Sandy clay
	Silty clay
	Clay
C HIGH	$2\frac{1}{2}$ in. or more per ft depth
	Loamy very fine sand
	Very fine sand
	Very fine sandy loam
	Peat
	Loamy peat
	Peaty loam

Soil Moisture Tension

The plant roots take up water from the film held around each soil particle by the force of surface tension and they must overcome this force in order to absorb the water. As the film remaining around the particles becomes thinner this force increases, and the roots must therefore exert an ever increasing suction. This suction may be measured (e.g., with a tensiometer, see page 61) and its magnitude is an indication of the dryness of the soil; at field capacity it is equivalent to about $\frac{1}{20}$th atmosphere and at permanent wilting point it is about 15 atmospheres. Although these forces are the same for all soils, the actual quantity of water needed to produce them differs in different soils. The curve relating the forces required to release water from a soil at different moisture contents is known as its moisture release characteristic curve and examples are shown in Fig. 2.

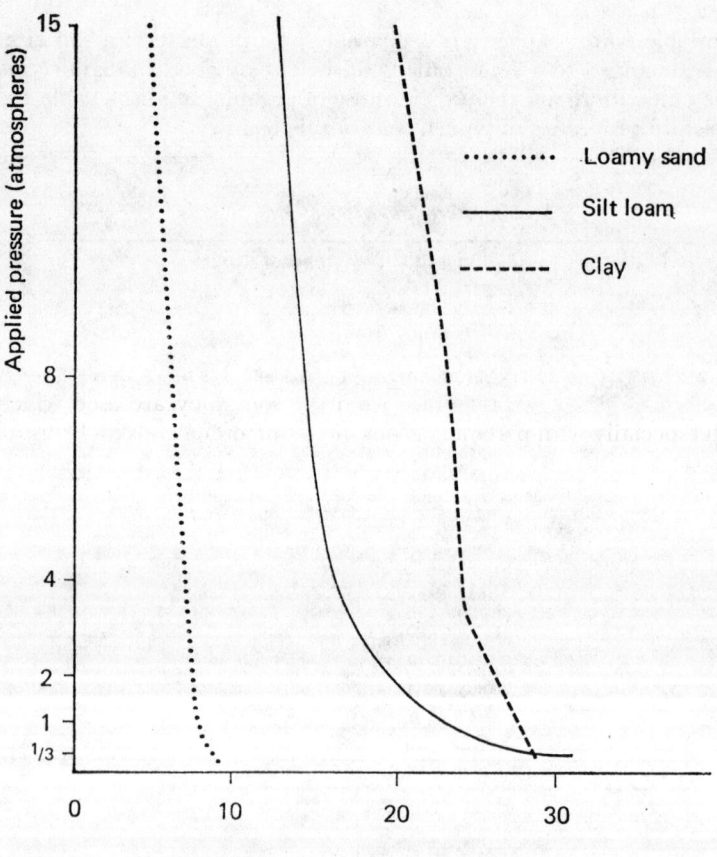

Fig. 2. Moisture movement curves

THE DETERMINATION OF SOIL MOISTURE PROPERTIES

Apparent Specific Gravity

Determined by driving hollow cylinders of known volume vertically into undisturbed platforms prepared at the required depths in a suitable pit. The soil samples are removed and weighed before and after drying at 105°C. Results are expressed as the ratio of weight of soil to the equivalent volume of water; values range from 0·5 (coarse sand) to 1·9 (silt loam).

Soil Moisture Content

Total soil moisture content (NOT available water content) may be determined gravimetrically by weighing a sample before and after drying to constant weight at 105°C. Moisture content may also be deduced from the readings of various instruments buried in the soil. For example, the neutron scattering method (neutron probe) measures the concentration of hydrogen ions in the soil; in mineral soils most of the hydrogen forms part of the soil water, but in organic soils some is a component of the organic matter so that in such soils expression of the results in absolute terms is complicated. The neutron scattering method is generally recognized as the best means for investigating the removal of water by roots from different depths over long periods, but the equipment is expensive and best suited to research.

Soil moisture status may be deduced from changes in related soil properties such as electrical resistance or capacitance or thermal conductivity, but again these methods are more suitable for research than for commercial farming.

The tensiometer consists essentially of a waterfilled porous pot buried in the soil; water moves into or out of the pot until the suction set up within equals the moisture tension in the surrounding soil (see page 4). The suction inside the pot may be measured with a manometer or Bourdon gauge. Tensiometers give a direct measurement of the stress which roots must overcome in order to extract water from the soil; they are used widely in research especially with perennial crops including orchard trees. Instruments suitable for commercial use are available, but their installation and interpretation of results require expert guidance.

Field Capacity

Best measured by gravimetric determination of the moisture content of samples taken from undisturbed soil after it has been completely saturated (either by rain or artificially) and allowed to drain naturally for at least two days, during which time evaporation from the surface has been prevented by means of a cover. Results are expressed as the amount of water present stated as a percentage of the dry soil; values range from 6·8 per cent (coarse sand) to 33 per cent (silt loam). Field capacity may also be determined by laboratory methods.

Permanent Wilting Point (or Percentage)

May be measured by growing sunflower seedlings in samples of the soil until these are fully pervaded with roots. The soil surface is then covered to prevent evaporation, and no more water is applied. When certain leaves of the plants wilt and do not recover when the culture is placed in a saturated atmosphere, the soil is at permanent wilting point and its moisture content is determined; values range from 1·8 per cent (coarse sand) to 13 per cent (silt loam). Wilting point may also be determined by laboratory methods.

Available Water Capacity

This is the amount of water held between the upper and lower limits (field capacity and permanent wilting point). It may be expressed for example in inches per foot depth or centimetres per metre depth calculated on a volume basis according to the formula:

$$\text{Available water capacity} = \frac{(\text{Field capacity} - \text{Wilting point}) \times \text{Apparent specific gravity}}{100}$$

To circumvent difficulties with units, available water capacity is better expressed as a dimensionless ratio or a decimal fraction. Thus available water capacity of $\frac{1}{12}$ or 0·83 can readily be converted to any units convenient for the particular circumstances and becomes 1 in. per 12 in. soil depth, 10 mm per 120 mm, 0·8 in. per 10 in. or 8 mm per 0·1 m.

Available water capacity ranges from 0·8 in. per foot (coarse sand) to 2·8 in. per foot (silt loam).

Moisture Release Characteristic Curve

Wet soil samples, standing upon a permeable membrane are subjected to constant gas pressure; when no more water is expressed from the sample, i.e., the force retaining the water equals the imposed force, the moisture content of the sample is determined. The curve relating such moisture contents to the appropriate pressures (tensions) is the moisture release characteristic curve (see Fig. 2); it illustrates the fact that the lower the soil moisture content, the greater the force which roots must overcome in order to extract water. Moisture release characteristics differ for different soils. In Fig. 3 the data are plotted in terms of available soil water; it is evident that the silt loam contains a large amount of available water which is released at comparatively low tensions, i.e., it is *readily* available, whereas the loamy sand contains less water of which a greater proportion is released only at higher tensions.

NOTE. The techniques described are not suitable for all soils; although they have been used successfully for agricultural soils ranging from sands to silt loams and peats they are not adequate for dealing with certain chalk soils. Further difficulty arises from the presence of abundant stones and suitable adjustments must be made.

Movement of Water in the Soil

For practical purposes water normally moves only downwards in the soil, that is, by drainage. Upward movement or sideways movement by capillarity from comparatively wet zones to dryer zones takes place only very slowly. Sideways movement can also take place by mass flow as when a ditch has

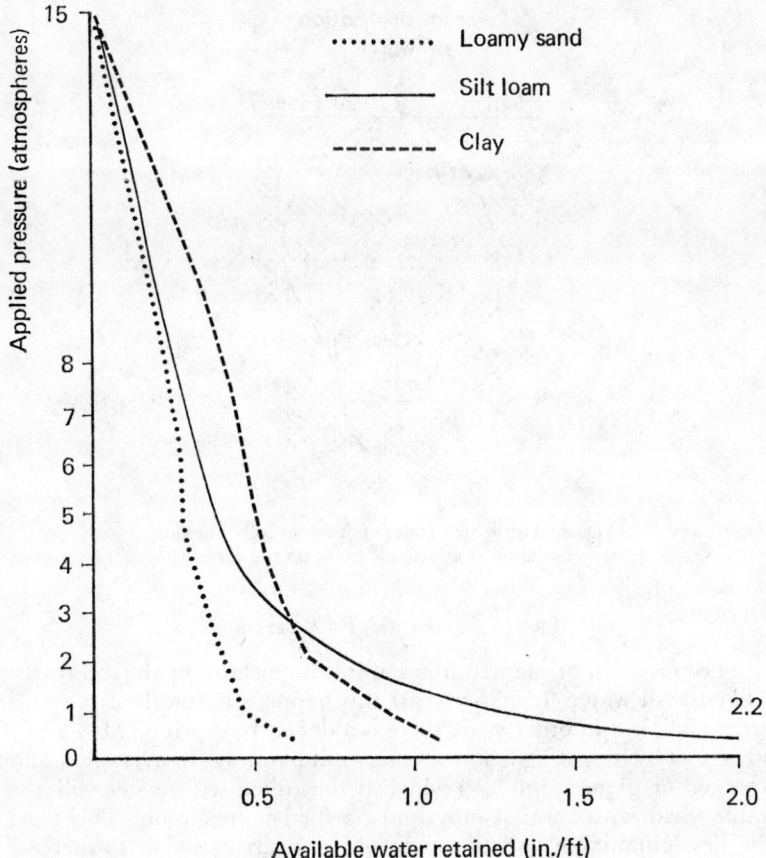

Fig. 3. Moisture release curves

been dammed and the water is forced to flow sideways saturating the soil as it goes.

When water falls on to the soil surface its subsequent movement within the soil is mainly downwards but there may also be some sideways movement. The relative extents of downward and sideways movement depend on soil type, as indicated in Fig. 4. This is the pattern of distribution of water which takes place, for example, in trickle irrigation or in the flow of water from the bottom of a channel in furrow irrigation.

When water is applied generally over the soil surface it first saturates the upper layers by filling the pore spaces and forcing out all the air. As more water falls on the surface this saturated air-less layer extends deeper until the whole of the soil is saturated and water flows into the subsoil or into the drains. As described previously, when the application of water ceases, downward water movement continues until the soil reaches the state of field capacity and then virtually stops. Thus the effect of adding different quantities of water to a soil is merely to alter the depth to which the soil is eventually brought to field capacity. It is not possible by ordinary means to 'half wet' the soil or to raise its moisture content to any level other than field capacity.

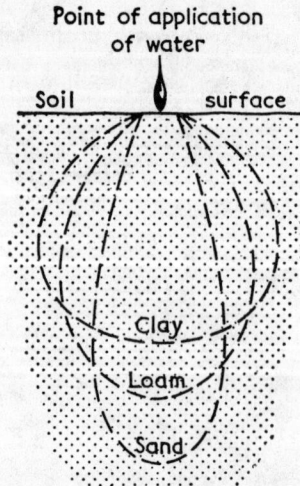

Fig. 4. The wetted zones of different soils when water is applied at a single point on the surface

IRRIGATION IN PRACTICE

Before a crop is sown or planted in a soil it is important for the soil reservoir to be filled with water. In many years this happens naturally as a result of winter rainfall, but in other years there is a deficit by April or May and this should be met by irrigating before sowing or at planting. In average summers a crop sown or planted into soil which is thus filled with water will give a profitable yield with rainfall only and no further irrigation. This yield is not the best obtainable and the main object of irrigation is to increase it substantially by relieving moisture stress at particular stages of crop growth, when the plants will respond by increasing the proportion of marketable parts. The second object of irrigation is to provide insurance against the comparatively few years in which rainfall is insufficient to give a profitable yield.

INFILTRATION RATE

Some soils can absorb water more rapidly than others, and when irrigating it is important to keep the rate of application slow enough for the soil to take up the water as fast as it falls. If the application rate exceeds the infiltration rate there will be sideways flow of water over the surface of soil; this causes erosion and also wastage of water. During irrigation the field should be inspected frequently and, if puddles and rivulets appear on the soil surface, watering should be stopped at once. Watering at a slower rate may be restarted when the puddles have disappeared.

The soil may also be damaged by excessive drop size during irrigation. The impact of large drops can cause the soil crumbs to lose their structure and run together, thus further reducing the rate at which the soil can take up water and also spoiling the tilth. One of the benefits of mulching the soil under irrigated crops is the reduction of this form of damage to soil structure. Fine drops or mist have a much less damaging effect on the soil surface and on some soils water may be applied at the very high rate of several inches per hour provided the drop size is no bigger than a heavy mist. Such small drops cannot, of course, be thrown as far as large drops, and so with conventional irrigation equipment there is a minimum drop size below which the equipment is ineffective in distributing the water adequately.

Moisture-Sensitive Growth Stages

It is generally accepted that plants grow to the maximum size possible under the prevailing light, temperature and nutrition if they are not subject to moisture stress at any time during their life. On the other hand, experiments have shown that with some plants this maximum size can still be achieved in spite of temporary moisture stress at certain times. For example peas can still give maximum yield even if they have been subject to drought before the flowers formed provided the plants are *not* subject to moisture stress at certain other critical times, i.e., at about the time of fertilization and pod swelling. This moisture sensitivity at flowering is found in many plants, ranging from cereals to tree fruits. Clearly, it is of paramount importance for the practical irrigator to ensure that the crop is not short of water at these particular growth stages, even if there has been stress at other times, and to irrigate when the moisture-sensitive stages are reached, if the weather is dry. The *amount* of this irrigation is not important and is unrelated to soil moisture status, but sufficient water must be supplied to relieve internal water stress—$\frac{1}{2}$ in. or more—at exactly the right stage of plant growth. This implies that careful and frequent observation of the plants themselves is required rather than assessment of soil moisture deficit.

Many important crop plants, such as the brassicae, do not have obvious moisture sensitive growth stages during their life span under commercial conditions (although they may have such stages, for example, after they have bolted). Experiments with such plants have shown that short term droughts of two or three weeks have their most damaging effect on yield when they occur either at the time of crop establishment, or during the last few weeks before harvest. Thus the irrigator should ensure that these crops are not short of water at these two stages.

It has previously been remarked that the soil is usually close to field capacity in the spring, i.e., at planting or sowing time, and only in exceptional years will irrigation be needed to fill the soil reservoir for the first of these drought-sensitive periods with this type of crop.

The drought sensitive period just before harvest is probably the most important time for commercial irrigation. In most seasons, there will be a substantial soil moisture deficit at this time (Table 1) and this will probably exceed the quantity of water which the average irrigator is able to apply. Therefore, as a general rule for an average year, 1 or 2 in. should be applied to any crop of this type (cabbage, cauliflower, kale and most kinds of roots), about a fortnight before harvest is due.

Artificial Reduction of Transpiration

The transpiration rate of plants may be reduced to some extent by changing meteorological factors such as sunshine, temperature and humidity in such a way as to reduce the evaporative power of the environment, or by spraying with substances which either obstruct the stomata or interfere with the internal metabolism of the plant.

Sheltering and shading are two possible means of modifying the environment; the use of mist has been suggested to obviate the effects of temporary wilting in hot sunshine, or to increase general humidity by repeated application throughout the day.

Anti-transpirant materials which obstruct the stomata include solutions of plastics in volatile solvents which can be sprayed on to the foliage; these are commercially available and used for preventing the wilting of cuttings, or of flowers, foliage and pot plants when displayed for sale.

Materials which interfere with internal plant processes are the subject of experimentation and not yet suitable for commercial use.

Mulching

As already mentioned, water loss can occur by evaporation from the soil surface so long as this is moist, and not sheltered by foliage. Depending upon the circumstances, up to $\frac{3}{4}$ in. can be lost in this way before the surface becomes so dry that evaporation virtually ceases. Cultivation which brings more moist soil to the surface will increase the water loss from the soil reservoir, but very light cultivation which retains a thin surface mulch of dry soil has been held to conserve soil moisture by preventing its upward movement from the soil below.

Many different materials may be applied to the surface to form sealing mulches, including waste foliage, leaves or straw, plastic film, or sprayed materials such as bitumen emulsion, rubber solution or other substances which form an impermeable film. Most of these materials are effective in reducing water loss and, with the exception of plastic sheeting will permit the ingress of rain or irrigation through cracks in the mulch, without undue hindrance. Plastic sheeting should be perforated at intervals to allow rain to reach the soil. Certain mulch materials, such as straw, take up nitrogen from the soil as they rot down, and this must be compensated for in the fertilizer applied to the crop.

The sprayed mulches are especially suitable for reducing water loss from the soil surface over sown seeds, thus aiding their germination and establishment. The seedlings have no difficulty in piercing the mulch from below; one disadvantage is that the mulch not only improves establishment of the crop seedlings, but also helps the growth of weed seedlings, which may be more difficult to deal with because of the mulch.

METEOROLOGICAL ASPECTS

The use of water by plants through transpiration is affected by meteorological factors; the supply of water to the soil for use by plants through their roots is controlled by the rainfall. It therefore follows that the liability of a plant to a check in growth because of the inadequacies of the water balance depends largely on the weather during the growing season. On the whole our weather often leads to moisture deficiencies in plants with a risk of reduced yield. The meteorological assessment of irrigation need aims at finding out how often this occurs and to what extent.

The Rainfall Income

There are over 6,000 official rainfall reporting stations in Britain, many of which have been in operation over a long period, producing a knowledge of the daily, monthly and annual rainfall climate which is second to none. The inaccuracies inherent in such data are probably less than 5 per cent. Nevertheless, knowledge of the past does not ensure foreknowledge of the future. All that can be said is that the annual rainfall of the next couple of decades is unlikely to differ greatly in level from that of the last 20 years, although the pattern sequence will be altered. This distribution of rain between the various months is likely to be more variable, but again the general picture will be roughly similar and, in any case, we have no access to other information regarding the future. The monthly averages of rainfall appropriate to the area in question are therefore the basis of any planning operation, provided that the variations about the average are taken into full consideration.

The Transpiration Expenditure

A good approximation of the potential transpiration of plants (that is to say the transpiration of a ground cover of a freely growing crop not subject to soil moisture check) can be found by calculation from meteorological measurements of sunshine (or radiation), temperature, humidity and wind. Full details are to be found in the Ministry's Technical Bulletin 16* *Potential Transpiration*.

Provided that account is taken of the proximity to the coast and to the height above sea level, the averages quoted in Technical Bulletin 16 are reasonably accurate, with local errors slightly greater than for rainfall, but of the same order and not misleading in practice. Extrapolation into the future is more justifiable than for rainfall, consequently future planning can be made dependent on past records with greater confidence.

Averages for each area in the United Kingdom are given in the Appendix.

The Water Balance

With knowledge of past income and expenditure, a water balance can be drawn up, on a daily, weekly, 10-day, 15-day or monthly basis. Transpiration data, obtained by calculation from meteorological data are rarely reliable on a daily basis; monthly balances can obscure the existence of temporary dry spells. Balances struck at 7, 10 or 15 day intervals are therefore preferable.

Two facts must be borne in mind: one is that the soil cannot be restored to more than capacity (except temporarily, see page 2), unless it is badly drained and becomes waterlogged—excess rainfall must therefore be regarded as lost to the drainage system once capacity has been reached; the other is that plants, when subject to a moisture stress, are likely to transpire at a rate below that of the potential. Under correct irrigation, however, neither of these states of affairs should occur, and an irrigation plan is designed with this object in mind.

---

* Technical Bulletin 16. *Potential Transpiration*, price 34p (by post 39½p), obtainable from H.M. Stationery Office or through any bookseller.

Consider for example a grass crop growing in a soil which is at field capacity at the end of March. If the potential transpiration during April is 2·0 in. and the rainfall is 3·0 in., the soil will still be at field capacity at the end of the month despite the excess of 1 in. of incoming rain over outgoing transpiration.

If during the next month, May, the potential transpiration is 3·5 in. and there is no rain at all, the potential soil moisture deficit at the end of the month will be 3·5 in. but the actual deficit will be more in the neighbourhood of 2·75 in., as the grass will not have been able to take up enough moisture from within its root range to meet the full transpiration demand and its growth will have suffered.

If, on the other hand 1·5 in. is added at the middle and again at the end of the month, the soil moisture deficit at the beginning of June will be 0·5 in. and full grass growth will have been maintained throughout.

Requirements of the Plant

Plant requirements are usually defined in terms of soil moisture deficit and these will vary with the type of crop and the stage of development. Irrigation plans to meet these requirements will vary with many factors such as type of soil, equipment and manpower available and potential economic returns.

Four typical plans are as follows:

PLAN 1   Soil restored to capacity when the soil moisture deficit reaches 1 in.

PLAN 2   Soil restored to capacity when the soil moisture deficit reaches 2 in.

PLAN 3   Deficit reduced to 1 in. when it reaches 3 in.

PLAN 4   Deficit reduced to 2 in. when it reaches 5 in.

Past Analyses

Agricultural meteorologists working with the Agricultural Development and Advisory Service have examined the weather in detail over 20 years for 80 sites in relation to the above four plans. Taking into account each possible period of 2, 3, 4, 5 or 6 months during the growing season, the following types of result were obtained:

the number of years in 20 when irrigation was needed;
the irrigation need in the fifth driest year in 20, in inches; and
the total irrigation need in 20 years in inches, adjusted so that the four driest years only receive the irrigation of the fifth driest. This adjustment was made in the 20 year totals because it may not be economical or practical to plan for a complete irrigation schedule in the driest years.

Frequency of Need Over Entire Growing Season

The results of this detailed investigation are published as a series of maps in the *Atlas of Long-term Irrigation Needs for England and Wales* (M.A.F.F. 1967)*. Graphical representations of some of the findings are reproduced in Appendix II, Figures 24, 27, 30 and 33. Given the average rainfall and the average potential transpiration for a specified place, it is possible to read off from these graphs the frequency of irrigation need for each of the four plans specified in the foregoing paragraph.

PLAN I AND PLAN 2

These plans allow for a maximum soil moisture deficit of 1 in. and 2 in. respectively. Over the farmland areas of the British Isles, as distinct from upland rough grazing, there are very few years when such deficits do not arise at some period during the growing season. The theoretical need for irrigation to maintain easily available soil moisture at all times from April to September was found to be 17 years or more out of 20, and in the drier parts of eastern England the frequency is 20 out of 20.

It must however be realized that Plan 1 presents a very stringent requirement, and is one which is only likely to be adopted for a few very valuable crops and over limited periods. Even so it will be only a small minority of years when such a plan need not be implemented (see Fig. 24, Appendix II).

Plan 2 applied throughout the growing season could be suitable for a crop such as grass, when irrigation is part of a husbandry system involving very intensive stocking or a continual maximum supply of grass or fodder for convervation by means of artificial drying. In such cases, a need frequency of 17 years or more out of 20 can be confidently assumed (see Fig. 27, Appendix II).

PLAN 3

This plan is less demanding, allowing deficits to increase to 3 in. and thereupon reducing them to 1 in. It is suitable for farm crops other than grass and for bush fruit. If it is decided to maintain this standard of soil moisture throughout the growing season, the frequencies can be deduced from Fig. 30, Appendix II.

To obtain an idea of what this tells us in regard to a specific farm, Table 2 has been prepared to show the critical average annual rainfalls in each county which would be equivalent to an irrigation need of 10, 9, 8, 7, 6, 5 and 4 years in 10.

Taking 7 years in 10 as a sensible requirement to make investment in irrigation practice an economic proposition, Table 2 shows that:

> In the East and South East, all areas have this frequency or more.
>
> In the East Midlands, all counties, except Derbyshire where the rainfall limit (for 7 years in 10) is 35 in.

---

* *Atlas of Long Term Irrigation Needs for England and Wales*, price £2·07½ (by post £2·49½) obtainable from the Ministry of Agriculture, Fisheries and Food (Publications), Tolcarne Drive, Pinner, Middlesex HA5 2DT.

In the West Midlands—all Warwickshire and Worcestershire, and with rainfall limits ranging from 34 in. in Staffordshire to 37 in. in Cheshire.

In the North East, the rainfall limits are 29·5 in. in Northumberland and Durham increasing to 33·5 in. in the East and West Ridings.

In the North West, the corresponding limits are 31 in. in Cumberland increasing to 38·5 in. in Lancashire, where high transpiration occurs in summer.

In the South West, where again transpiration is high, the rainfall limits vary from 38 in. in Gloucestershire to 46 in. in Cornwall.

In Wales, some counties have no areas which attain the 7 in 10 requirement, namely Cardigan, Carmarthen and Merioneth. Only small areas in this relatively wet part of the British Isles exist elsewhere in Wales; the rainfall limit ranges from 34·5 in. in Flint to 42 in. in Pembroke.

PLAN 4

This is the least demanding schedule of the four and was chiefly designed for tree fruit crops.

Table 3 has been prepared from Fig. 33 in the same manner as Table 2. It shows the limiting average rainfall in the various counties of England and Wales for irrigation needs for Plan 4, of frequencies 9, 8, 7, 6, 5 and 4 years out of 10.

Again assuming 7 years in 10 as a significant frequency it will be seen that practically no area in Wales or the North attains this requirement, but that practically all Eastern England still falls within the category.

Requirements Over Limited Periods

The high frequencies of need deduced in this analysis are partly because a very strict maintenance of available soil moisture over a six month period has been stipulated.

In practice, as will be seen later, it is often possible to get worthwhile results by irrigation of periods of shorter duration. An attempt has been made in the Appendix to give specimen frequencies of irrigation need of such a limited type. Details are given for regional areas*, including needs in Scotland and Northern Ireland where the climate is such that a full six months cover of irrigation is rarely necessary in most areas.

It is, however, impossible to cover all specific purposes, and although the diagrams in this Bulletin enable such calculations to be made, it is always wise to consult the A.D.A.S. advisers to make adequate assessments and consequent plans.

---

* The counties referred to throughout this bulletin are the old administrative ones as existing in 1973.

Table 2

Limits of annual rainfall (in.) for various frequencies of irrigation need (years in 10) to carry out plan 3 from April to September

Region	County	\	\	\	Frequency of need	\	\	\
		10	9	8	7	6	5	4
North	Cumberland	none	none	none	31.0	34.5	38.5	43.0
	Durham	none	24.0	27.0	29.5	32.0	35.0	38.5
	North Riding	none	26.0	29.0	32.5	35.0	39.0	43.5
	Northumberland	none	none	27.0	29.5	32.0	34.5	37.5
	Westmorland	none	none	none	35.0	38.5	43.0	48.0
Yorkshire and Lancashire	Lancashire	none	none	35.0	38.5	42.0	46.0	51.0
	East and West Riding	none	26.5	30.0	33.5	37.0	41.0	45.0
West Midlands	Cheshire	none	30.0	33.5	37.0	40.5	44.5	48.5
	Hereford	none	29.0	32.5	36.0	39.5	43.5	48.0
	Shropshire	none	27.5	31.0	34.5	38.0	all	
	Staffordshire	none	27.5	31.0	34.0	37.0	41.0	44.5
	Warwickshire	none	29.0	all				
	Worcestershire	none	30.0	all				
East Midlands	Derbyshire	none	28.0	32.0	35.0	38.5	42.5	47.5
	Kesteven, Lindsey	none	29.0	all				
	Leicestershire	none	28.5	all				
	Northamptonshire and Nottinghamshire	none	29.0	all				
	Rutland	none	28.0	all				
East	Bedfordshire	none	all					
	Cambridgeshire	21.5	all					
	Essex	24.5	all					
	Hertfordshire	none	30.5	all				
	Holland	none	all					
	Huntingdonshire	22.0	all					
	Norfolk and Suffolk	24.0	all					
South East	Berkshire	none	31.0	34.0	all			
	Buckinghamshire	none	30.0	33.0	all			
	Hampshire	none	35.0	38.0	all			
	Isle of Wight	none	all					
	Kent	26.0	35.0	all				
	Oxfordshire	none	30.0	33.0	all			
	Surrey	24.5	33.0	all				
	Sussex	none	35.0	all				
South West	Cornwall	none	38.0	42.0	46.0	50.0	55.0	60.0
	Devon and Dorsetshire	none	36.5	40.0	44.0	48.5	53.5	58.5
	Gloucestershire	none	31.0	34.5	38.0	all		
	Isles of Scilly	none	all					
	Somersetshire	none	32.0	36.0	40.0	44.0	49.0	54.0
	Wiltshire	none	32.0	36.0	39.5	43.5	all	

TABLE 2—continued

Region	County	Frequency of need						
		10	9	8	7	6	5	4
Wales	Anglesey	none	none	36·5	40·0	43·5	all	
	Brecon	none	none	35·0	38·0	41·0	45·5	50·0
	Caernarvon	none	none	35·0	38·0	42·0	46·0	51·0
	Cardigan	none	none	none	none	43·0	47·0	52·0
	Carmarthen	none	none	none	none	46·0	50·5	55·5
	Denbigh	none	30·0	33·5	36·5	40·0	44·0	48·5
	Flint	none	none	31·5	34·5	37·5	41·5	45·5
	Glamorgan	none	none	35·5	39·0	43·0	47·5	52·0
	Merioneth	none	none	none	none	none	44·0	48·5
	Monmouth	none	33·0	36·5	40·0	44·0	48·5	53·0
	Montgomery	none	30·0	33·5	36·5	40·0	43·5	48·5
	Pembroke	none	none	none	42·0	46·0	51·0	55·0
	Radnor	none	none	35·0	38·0	41·0	45·5	50·0

Note: The entry 'none' implies that no area of the country has this frequency.
The entry 'all' implies that all the areas of the country have this frequency.

TABLE 3

Limits of annual rainfall (in.) for various frequencies of irrigation need (years in 10) to carry out plan 4 from April to September

Region	County	Frequency of need					
		9	8	7	6	5	4
North	Cumberland						none
	Durham	none	none	none	25·0	27·0	29·0
	North Riding	none	none	none	27·0	29·5	32·0
	Northumberland	none	none	none	none	27·0	29·0
	Westmorland	none	none	none	none	none	34·5
Yorkshire and Lancashire	Lancashire	none	none	none	33·5	36·0	38·5
	East and West Riding	none	none	25·0	28·0	30·5	33·0
West Midlands	Cheshire	none	none	none	31·5	34·0	36·5
	Herefordshire	none	none	27·0	30·5	33·0	36·0
	Shropshire	none	none	none	29·0	31·5	34·0
	Staffordshire	none	none	26·0	29·0	31·0	33·5
	Warwickshire	none	25·0	27·5	all		
	Worcestershire	none	26·0	28·5	all		
East Midlands	Derbyshire	none	none	26·5	30·0	32·5	35·0
	Kesteven and Lindsey	none	25·5	all			
	Leicestershire	none	none	27·0	all		
	Northamptonshire and Nottinghamshire	none	25·5	28·0	all		
	Rutland	none	24·0	26·5	all		

TABLE 3—continued

Region	County	Frequency of need					
		9	8	7	6	5	4
East	Bedfordshire	none	all				
	Cambridgeshire	22·0	all				
	Essex	all					
	Hertfordshire	none	26·5	29·5	all		
	Holland	22·0	25·5	all			
	Huntingdonshire	22·5	24·0	all			
	Norfolk and Suffolk	23·5	27·5	all			
South East	Berkshire	none	26·5	29·0	30·5	32·5	all
	Buckinghamshire	none	26·0	28·5	31·5	all	
	Hampshire	27·0	30·0	32·5	34·0	35·5	all
	Isle of Wight	30·0	all				
	Kent	27·0	31·0	34·0	all		
	Oxfordshire	none	26·0	28·5	31·5	all	
	Surrey	25·5	29·5	32·5	all		
	Sussex	27·5	30·5	33·5	36·5	all	
South West	Cornwall	none	none	36·5	40·5	43·5	46·5
	Devon and Dorsetshire	none	31·0	34·5	38·0	41·5	44·5
	Gloucestershire	none	27·0	30·0	33·0	35·5	38·0
	Isles of Scilly	none	all				
	Somersetshire	none	none	31·5	35·0	37·5	40·0
	Wiltshire	none	none	31·0	34·5	37·0	40·0
Wales	Anglesey	none	none	none	34·5	37·0	39·5
	Brecon	none	none	none	none	none	37·0
	Caernarvon	none	none	none	none	35·0	37·5
	Cardigan	none	none	none	none	none	none
	Carmarthen	none	none	none	none	none	none
	Denbigh	none	none	28·0	31·5	34·0	36·0
	Flint	none	none	none	30·0	32·0	34·0
	Glamorgan	none	none	none	none	36·5	39·0
	Merioneth	none	none	none	none	none	none
	Monmouth	none	none	none	34·5	37·0	40·0
	Montgomery	none	none	none	31·0	33·5	36·0
	Pembroke	none	none	none	none	none	42·0
	Radnor	none	none	none	none	none	37·0

Planning Irrigation

WHEN considering whether to include irrigation in the farm enterprise, the following questions need to be answered:

Which crops will respond to irrigation?

How often will weather make this profitable?

How much water will be needed to obtain maximum yields?

What sources of water are available?

What legal problems are involved?

If water amounts are restricted, what alternative irrigation practices will give reasonable results?

What kind and quantity of equipment are needed?

What additional uses can be found for this equipment?

CROP RESPONSE TO IRRIGATION

Obviously all crops grown under permanent or temporary protection, including glasshouses, structures, cloches, tunnels, frames and the like, require irrigation. The special equipment and methods involved are discussed on pages 60–73.

The preceding pages have shown that the majority of crops grown in the open experience soil moisture deficits in most years. These may be due to short dry spells whose effects are unrecognized or underestimated but which can reduce yields. How far it is economically sound to correct these deficits by irrigation depends upon their frequency and the value of the crop. There is thus a differential scale of assessment from high-value crops such as strawberries and new potatoes to low-value crops such as grass. Nevertheless, the intrinsic value of the crop may not be the only consideration. For example, when a reliable grass supply forms an integral part of an intensive animal stocking system, the complete assurance of animal feed is important.

With permanent crops such as tree and bush fruits, judicious irrigation improves the produce not only in the current year but also in succeeding years.

The establishment of spring sown or transplanted crops can be aided by irrigation, and even cereals may benefit from watering when a dry spell follows spring drilling.

How Often Will Irrigation be Profitable?

The number of years when irrigation is required depends upon the crop, the climate and, to a less extent, on soil type. Crop needs are discussed on pages 73–97 (individual crop requirements). The implication of climate, which is related to the geographical location of the farm, has already been described in the previous Section in respect of various watering plans. Soil type can influence the choice of plan and its implementation, for example, less retentive soils will require more frequent watering. It is important to appreciate that water use is not *greater* on the poorly retentive soils, but these have a smaller soil moisture reservoir than the better soils and therefore require more frequent replenishment. (See Section 1).

How Much Water is Required for Maximum Yield?

As long as there is adequate available water in the soil, transpiration by plants and evaporation from the soil surface are largely controlled by the weather, as explained in sections 1 and 2. Therefore the long-term irrigation needs can be estimated from climatic data for the area. These meteorological aspects will now be discussed in greater detail.

Maximum Seasonal Water Need

The greatest need for water will occur during the driest summer when irrigation is planned for the whole season. It is not generally practical to plan for the extreme year, and as explained earlier, the detailed four-plan analysis extracted the needs for the 5th driest year in 20. The results are illustrated in Figures 25 (Plan 1), 28 (Plan 2), 31 (Plan 3) and 34 (Plan 4), Appendix II.

All that is needed to use these diagrams is knowledge of the average rainfall and average potential transpiration over the period concerned (in this seasonal case, 6 months). The difference between transpiration and rainfall establishes a point on the bottom horizontal scale; moving upwards vertically from this point to the 6 month curve, thence horizontally to the scale on left establishes the maximum need for 16 years out of the same 20 (the 'adjusted' need).

Table 4 has been prepared to show results obtained in this manner for Plans 2, 3 and 4 for each County in England and Wales over a variety of average annual rainfall. Counties in Scotland and N. Ireland have not been included as a 6 month irrigation schedule is rarely necessary in those areas.

Details have not been included for Plan 1 because it is not likely to be operative over the entire season.

The amounts of water required in these dry years can be extremely high in low rainfall areas. In a very dry year such amounts of water may not be available unless adequate local storage has been planned.

The maximum water needs over shorter periods (of less than 6 months) can be read off the diagrams using the appropriate rainfall and transpiration figures. A selection of results for all areas of the British Isles is given in the Appendix.

Table 4

Maximum seasonal (adjusted) need in 20 years (in.)

Region	County	Plan	\multicolumn{7}{c}{Average annual rainfall (in.)}						
			21	23	25	27	30	35	40
North	Cumberland	2	—	—	—	—	—	4·5	3·9
		3	—	—	—	—	—	3·6	3·1
	Durham	2	—	—	7·3	6·4	5·4	4·4	3·7
		3	—	—	6·4	5·5	4·6	3·5	2·9
		4	—	—	4·4	4·0	3·5	2·8	—
	North Riding	2	—	—	7·8	7·1	6·1	4·9	4·1
		3	—	—	6·9	6·2	5·2	4·0	3·3
		4	—	—	4·8	4·3	3·8	3·0	—
	Northumberland	2	—	—	—	6·3	5·4	4·4	3·8
		3	—	—	—	5·3	4·6	3·4	3·0
		4	—	—	—	3·9	3·5	2·8	—
	Westmorland	2	—	—	—	—	—	5·4	4·5
		3	—	—	—	—	—	4·3	3·6
		4	—	—	—	—	—	3·5	—
Yorkshire and Lancashire	Lancashire	2	—	—	—	—	—	5·8	4·7
		3	—	—	—	—	—	4·9	3·8
		4	—	—	—	—	—	3·7	3·0
	East and West Riding	2	—	9·5	8·4	7·5	6·4	5·1	4·2
		3	—	8·6	7·6	6·7	5·4	4·2	3·4
		4	—	6·6	5·4	4·6	4·0	3·3	—
West Midlands	Cheshire	2	—	—	—	—	7·2	5·4	4·5
		3	—	—	—	—	6·3	4·6	3·6
		4	—	—	—	—	4·4	3·5	2·9
	Hereford	2	—	—	9·4	8·2	7·0	5·3	4·5
		3	—	—	8·5	7·4	6·1	4·5	3·6
		4	—	—	6·4	5·2	4·3	3·5	—
	Shropshire	2	—	—	—	7·4	6·6	5·2	4·4
		3	—	—	—	6·5	5·5	4·3	3·5
		4	—	—	—	4·5	4·0	3·4	—
	Stafford	2	—	—	—	7·5	6·2	4·9	4·0
		3	—	—	—	6·6	5·2	4·0	3·2
		4	—	—	—	4·6	3·9	3·0	—
	Warwick	2	—	—	9·2	8·0	6·7	—	—
		3	—	—	8·3	7·2	5·7	—	—
		4	—	—	6·2	5·0	4·1	—	—
	Worcester	2	—	—	10·2	8·7	7·3	—	—
		3	—	—	9·2	7·8	6·4	—	—
		4	—	—	7·2	5·6	4·5	—	—

TABLE 4—*continued*

Region	County	Plan	Average annual rainfall (in.)						
			21	23	25	27	30	35	40
East Midlands	Derby	2	—	—	—	7·8	6·8	5·3	4·4
		3	—	—	—	7·0	5·8	4·4	3·5
		4	—	—	—	4·9	4·1	3·4	—
	Kesteven and Lindsey	2	—	11·2	10·1	9·1	—	—	—
		3	—	10·2	9·1	8·2	—	—	—
		4	—	8·4	7·2	6·1	—	—	—
	Leicester	2	—	10·5	9·1	7·8	6·5	—	—
		3	—	9·5	8·2	7·0	5·5	—	—
		4	—	7·7	6·1	4·8	4·0	—	—
	Northampton and Nottingham	2	—	10·9	9·4	8·1	6·7	—	—
		3	—	10·0	8·5	7·3	5·7	—	—
		4	—	8·2	6·5	5·1	4·1	—	—
	Rutland	2	—	10·3	9·8	7·8	—	—	—
		3	—	9·4	8·0	6·9	—	—	—
		4	—	7·3	5·8	4·8	—	—	—
East	Bedford	2	—	11·6	10·1	—	—	—	—
		3	—	10·6	9·1	—	—	—	—
		4	—	8·9	7·2	—	—	—	—
	Cambridge	2	13·0	11·2	—	—	—	—	—
		3	12·0	10·2	—	—	—	—	—
		4	10·3	8·4	—	—	—	—	—
	Essex	2	15·0	13·5	11·4	—	—	—	—
		3	14·0	12·4	10·4	—	—	—	—
		4	12·5	10·8	8·5	—	—	—	—
	Hertford	2	—	—	10·3	9·1	7·4	—	—
		3	—	—	9·4	8·2	6·5	—	—
		4	—	—	7·3	6·1	4·5	—	—
	Holland	2	—	11·3	9·6	8·4	—	—	—
		3	—	10·3	8·7	7·6	—	—	—
		4	—	8·5	6·7	5·4	—	—	—
	Huntingdon	2	13·0	11·3	—	—	—	—	—
		3	12·0	10·3	—	—	—	—	—
		4	10·3	8·5	—	—	—	—	—
	Norfolk and Suffolk	2	14·0	12·3	11·0	10·0	—	—	—
		3	13·0	11·2	10·0	9·0	—	—	—
		4	10·5	9·6	8·3	7·0	—	—	—
South East	Berkshire	2	—	—	10·5	9·0	7·4	5·6	—
		3	—	—	9·5	8·0	6·6	4·7	—
		4	—	—	7·7	6·0	4·6	3·6	—

TABLE 4—continued

Region	County	Plan	Average annual rainfall (in.)						
			21	23	25	27	30	35	40
	Buckingham	2	—	—	10·0	8·6	7·1	—	—
		3	—	—	9·0	7·8	6·2	—	—
		4	—	—	7·0	5·6	4·3	—	—
	Hampshire	2	—	—	—	12·1	9·6	6·9	5·2
		3	—	—	—	11·1	8·6	6·0	4·3
		4	—	—	—	9·5	6·6	4·2	3·4
	Isle of Wight	2	—	—	—	—	11·4	8·5	—
		3	—	—	—	—	10·4	7·7	—
		4	—	—	—	—	8·6	5·5	—
	Kent	2	15·0	14·0	13·0	11·3	9·3	6·9	—
		3	14·0	13·0	12·0	10·3	8·4	6·0	—
		4	11·5	10·8	10·0	8·5	6·3	4·2	—
	Oxford	2	—	—	9·9	8·5	7·0	—	—
		3	—	—	8·9	7·7	6·1	—	—
		4	—	—	6·9	5·5	4·2	—	—
	Surrey	2	—	14·0	12·0	10·5	8·6	6·5	—
		3	—	13·0	11·0	9·5	7·8	5·5	—
		4	—	10·8	9·5	7·7	5·6	4·5	—
	Sussex	2	—	—	—	11·7	9·1	6·5	—
		3	—	—	—	10·7	8·2	5·5	—
		4	—	—	—	9·0	6·1	4·5	—
South West	Cornwall	2	—	—	—	—	—	8·1	6·3
		3	—	—	—	—	—	7·3	5·3
		4	—	—	—	—	—	5·1	3·9
	Devon and Dorset	2	—	—	—	—	10·0	7·5	5·9
		3	—	—	—	—	9·0	6·7	4·9
		4	—	—	—	—	7·0	4·6	3·7
	Gloucester	2	—	—	—	9·0	7·4	5·6	—
		3	—	—	—	8·0	6·6	4·7	—
		4	—	—	—	6·0	4·6	3·6	—
	Scillies	2					10·8		
		3					9·8		
		4					8·0		
	Somerset	2	—	—	—	—	8·0	6·3	5·2
		3	—	—	—	—	7·2	5·3	4·3
		4	—	—	—	—	5·0	3·9	3·3
	Wiltshire	2	—	—	—	—	8·1	6·2	4·9
		3	—	—	—	—	7·3	5·2	4·0
		4	—	—	—	—	5·1	3·9	3·1

TABLE 4—*continued*

Region	County	Plan	Average annual rainfall (in.)						
			21	23	25	27	30	35	40
Wales	Anglesey	2	—	—	—	—	—	6·5	5·7
		3	—	—	—	—	—	5·5	4·8
		4	—	—	—	—	—	4·0	3·6
	Brecon	2	—	—	—	—	—	6·0	4·9
		3	—	—	—	—	—	5·0	4·0
		4	—	—	—	—	—	3·8	3·0
	Caernarvon	2	—	—	—	—	—	6·1	4·9
		3	—	—	—	—	—	5·1	4·0
		4	—	—	—	—	—	3·9	3·0
	Cardigan	2	—	—	—	—	—	—	5·1
		3	—	—	—	—	—	—	4·2
		4	—	—	—	—	—	—	3·3
	Carmarthen	2	—	—	—	—	—	—	5·3
		3	—	—	—	—	—	—	4·4
		4	—	—	—	—	—	—	3·4
	Denbigh	2	—	—	—	8·6	7·1	5·5	4·5
		3	—	—	—	7·8	6·2	4·6	3·6
		4	—	—	—	5·6	4·3	3·5	3·0
	Flint	2	—	—	—	7·8	6·5	5·1	4·1
		3	—	—	—	7·0	5·5	4·2	3·3
		4	—	—	—	4·9	4·0	3·3	—
	Glamorgan	2	—	—	—	—	—	6·2	4·9
		3	—	—	—	—	—	5·2	4·0
		4	—	—	—	—	—	4·0	3·0
	Merioneth	2	—	—	—	—	—	—	4·6
		3	—	—	—	—	—	—	3·7
		4	—	—	—	—	—	—	—
	Monmouth	2	—	—	—	—	—	6·5	5·0
		3	—	—	—	—	—	5·5	4·1
		4	—	—	—	—	—	4·0	3·2
	Montgomery	2	—	—	—	—	7·2	5·5	4·5
		3	—	—	—	—	6·3	4·6	3·6
		4	—	—	—	—	4·4	3·6	3·0
	Pembroke	2	—	—	—	—	—	—	5·4
		3	—	—	—	—	—	—	4·5
		4	—	—	—	—	—	—	3·5
	Radnor	2	—	—	—	—	—	6·1	5·1
		3	—	—	—	—	—	5·1	4·2
		4	—	—	—	—	—	3·9	3·3

Total Seasonal Need over 20 Years

To obtain an idea of average working costs, the average annual need or the total need of water over a period of years is required.

To obtain this figure, the detailed analysis assumed that the maximum water applied in any year was the calculated amount for the 5th driest in 20, (the adjusted need). This amount was assumed to be the amount used in the 4 drier years. The results are shown in Appendix II in Figures 25 (Plan 1), 28 (Plan 2), 31 (Plan 3) and 34 (Plan 4).

The procedure is the same as in finding the maximum (adjusted) need in any one year and the total (adjusted) need in 20 years can be found from knowledge of the average rainfall and average transpiration over the period in question.

The results for a 6 month programme are tabulated in Table 5, the total (adjusted) need for 20 years being given (to the nearest 5 in.) for each county in England and Wales for a variety of average annual rainfalls. The average annual usage is thus found by dividing by 20.

Average Water Needs Over Shorter Periods

The diagrams can be used to find the 20 year (adjusted) needs over periods of 2, 3, 4 or 5 months. A selection of such needs is given in the Appendix for all areas of the British Isles.

Alternative Plans or Schedules

Irrigation schedules other than those specified in Plans 1, 2, 3 or 4 cannot be quantified by means of these diagrams or tables. If information of this nature is required, advice should be sought.

Table 5

Total seasonal (adjusted) need in 20 years (in.)

Region	County	Plan	Average annual rainfall (in.)						
			21	23	25	27	30	35	40
North	Cumberland	2	—	—	—	—	—	50	40
		3	—	—	—	—	—	30	20
	Durham	2	—	—	95	85	70	50	40
		3	—	—	75	65	45	30	20
		4	—	—	50	40	25	15	—
	North Riding	2	—	—	105	90	80	60	45
		3	—	—	85	70	60	40	25
		4	—	—	60	45	35	20	—
	Northumberland	2	—	—	—	80	65	50	40
		3	—	—	—	60	45	30	20
		4	—	—	—	40	25	15	—
	Westmorland	2	—	—	—	—	—	65	50
		3	—	—	—	—	—	45	30
		4	—	—	—	—	—	25	15

TABLE 5—continued

Region	County	Plan	Average annual rainfall (in.)						
			21	23	25	27	30	35	40
Yorkshire and Lancashire	Lancashire	2	—	—	—	—	—	75	50
		3	—	—	—	—	—	55	35
		4	—	—	—	—	—	30	20
	East and West Riding	2	—	125	115	100	85	60	45
		3	—	105	95	80	60	40	30
		4	—	80	65	55	40	20	—
West Midlands	Cheshire	2	—	—	—	—	95	65	50
		3	—	—	—	—	75	45	30
		4	—	—	—	—	50	25	15
	Hereford	2	—	—	125	110	90	70	50
		3	—	—	105	90	70	50	30
		4	—	—	80	60	45	30	15
	Shropshire	2	—	—	—	100	85	65	50
		3	—	—	—	80	65	45	30
		4	—	—	—	55	40	25	15
	Stafford	2	—	—	—	100	80	55	45
		3	—	—	—	80	60	35	25
		4	—	—	—	50	35	20	—
	Warwick	2	—	—	125	105	85	—	—
		3	—	—	105	85	65	—	—
		4	—	—	75	60	40	—	—
	Worcester	2	—	—	135	115	95	—	—
		3	—	—	115	95	75	—	—
		4	—	—	90	65	50	—	—
East Midlands	Derby	2	—	—	—	105	90	65	50
		3	—	—	—	85	70	45	30
		4	—	—	—	55	45	25	15
	Kesteven and Lindsey	2	—	160	135	120	—	—	—
		3	—	125	110	100	—	—	—
		4	—	105	90	75	—	—	—
	Leicester	2	—	145	120	105	85	—	—
		3	—	115	100	85	65	—	—
		4	—	95	75	60	40	—	—
	Northampton and Nottingham	2	—	150	125	110	90	—	—
		3	—	120	105	90	70	—	—
		4	—	100	80	60	45	—	—
	Rutland	2	—	140	120	100	—	—	—
		3	—	115	100	85	—	—	—
		4	—	90	70	55	—	—	—

TABLE 5—*continued*

Region	County	Plan	Average annual rainfall (in.)						
			21	23	25	27	30	35	40
East	Bedford	2	—	160	135	—	—	—	—
		3	—	130	110	—	—	—	—
		4	—	110	90	—	—	—	—
	Cambridge	2	185	155	—	—	—	—	—
		3	150	130	—	—	—	—	—
		4	135	105	—	—	—	—	—
	Essex	2	220	190	155	—	—	—	—
		3	170	145	130	—	—	—	—
		4	150	130	105	—	—	—	—
	Hertford	2	—	—	140	120	100	—	—
		3	—	—	115	100	75	—	—
		4	—	—	95	75	50	—	—
	Holland	2	—	155	130	110	—	—	—
		3	—	125	110	95	—	—	—
		4	—	105	80	65	—	—	—
	Huntingdon	2	185	155	—	—	—	—	—
		3	150	130	—	—	—	—	—
		4	135	105	—	—	—	—	—
	Norfolk and Suffolk	2	195	175	150	135	—	—	—
		3	155	135	120	110	—	—	—
		4	140	120	100	85	—	—	—
South East	Berkshire	2	—	—	145	120	100	70	—
		3	—	—	115	100	80	50	—
		4	—	—	95	75	55	30	—
	Buckingham	2	—	—	135	115	95	—	—
		3	—	—	110	95	75	—	—
		4	—	—	90	70	50	—	—
	Hampshire	2	—	—	—	165	130	90	65
		3	—	—	—	130	110	70	45
		4	—	—	—	115	85	45	25
	Isle of Wight	2	—	—	—	—	160	115	—
		3	—	—	—	—	130	95	—
		4	—	—	—	—	110	70	—
	Kent	2	230	210	185	160	125	90	—
		3	180	160	145	130	105	70	—
		4	160	150	130	110	75	45	—
	Oxford	2	—	—	135	115	90	—	—
		3	—	—	110	95	70	—	—
		4	—	—	90	65	45	—	—

TABLE 5—*continued*

Region	County	Plan	Average annual rainfall (in.)						
			21	23	25	27	30	35	40
	Surrey	2	—	195	165	140	115	85	—
		3	—	150	130	115	95	65	—
		4	—	135	120	95	70	40	—
	Sussex	2	—	—	—	160	125	85	—
		3	—	—	—	130	105	65	—
		4	—	—	—	115	75	40	—
South West	Cornwall	2	—	—	—	—	—	110	85
		3	—	—	—	—	—	90	65
		4	—	—	—	—	—	60	40
	Devon and Dorset	2	—	—	—	—	135	100	75
		3	—	—	—	—	110	80	55
		4	—	—	—	—	90	55	35
	Gloucester	2	—	—	—	120	100	70	—
		3	—	—	—	100	80	50	—
		4	—	—	—	75	55	30	—
	Scillies	2					150		
		3					120		
		4					100		
	Somerset	2	—	—	—	—	110	85	60
		3	—	—	—	—	90	60	40
		4	—	—	—	—	60	40	20
	Wiltshire	2	—	—	—	—	110	80	55
		3	—	—	—	—	90	60	40
		4	—	—	—	—	60	35	20
Wales	Anglesey	2	—	—	—	—	—	85	70
		3	—	—	—	—	—	65	50
		4	—	—	—	—	—	40	30
	Brecon	2	—	—	—	—	—	75	55
		3	—	—	—	—	—	55	35
		4	—	—	—	—	—	35	20
	Caernarvon	2	—	—	—	—	—	80	55
		3	—	—	—	—	—	55	35
		4	—	—	—	—	—	35	20
	Cardigan	2	—	—	—	—	—	—	60
		3	—	—	—	—	—	—	40
		4	—	—	—	—	—	—	20
	Carmarthen	2	—	—	—	—	—	—	65
		3	—	—	—	—	—	—	45
		4	—	—	—	—	—	—	25

TABLE 5—*continued*

Region	County	Plan	Average annual rainfall (in.)						
			21	23	25	27	30	35	40
	Denbigh	2	—	—	—	115	95	70	50
		3	—	—	—	95	75	50	35
		4	—	—	—	70	50	30	20
	Flint	2	—	—	—	105	85	60	45
		3	—	—	—	85	65	40	30
		4	—	—	—	60	40	20	15
	Glamorgan	2	—	—	—	—	—	80	55
		3	—	—	—	—	—	60	35
		4	—	—	—	—	—	35	20
	Merioneth	2	—	—	—	—	—	—	50
		3	—	—	—	—	—	—	35
		4	—	—	—	—	—	—	20
	Monmouth	2	—	—	—	—	—	85	60
		3	—	—	—	—	—	65	40
		4	—	—	—	—	—	40	20
	Montgomery	2	—	—	—	—	95	70	50
		3	—	—	—	—	75	50	35
		4	—	—	—	—	50	30	20
	Pembroke	2	—	—	—	—	—	—	70
		3	—	—	—	—	—	—	50
		4	—	—	—	—	—	—	30
	Radnor	2	—	—	—	—	—	80	60
		3	—	—	—	—	—	55	40
		4	—	—	—	—	—	35	20

WATER SUPPLY—SOURCES

An adequate and dependable water supply of suitable quality is essential for any irrigation project. Irrigation water may be obtained from surface sources—rivers, streams and lakes, from ground water sources by means of wells or boreholes, or from a public supply. The suitability of each possible source needs to be carefully assessed and the most dependable supply chosen before irrigation equipment is installed.

Large amounts of water may be required for irrigation. It requires approximately 23,000 gal to cover 1 acre with 1 in. and a typical summer demand for a crop is 1 in. every 10 days, rising to a peak demand of 1 in. every 8 days. Equipment capable of applying 1 in. of water every 10 days to 30 or 40 acres is fairly common on a farm scale, and during drought spells will require a supply of some 100,000 gal per day.

A licence from the River Authority is required before the water can be legally abstracted from a surface or ground water source for spray irrigation, unless the area has been exempted from licensing control by an order under Section 25 of the Water Resources Act 1963. This applies whatever the quantity of water available. The licence will state the quantity of water authorized to be abstracted during the period or periods specified. In addition, the River Authority levies annual charges on all licence holders. Further information concerning licences and charges is given under the next heading, 'Licences'.

The adequacy of a water source depends on the total amount of water available during the irrigation season when compared with the requirement during that period. Any deficiency must be met either from storage, which must be filled outside the irrigation season, or from some other source. Even where the total flow during the season is adequate, storage may be needed to cope with peak rates of demand during the season. The increasing demand for water from all purposes means that less and less water is available during the summer for new or increased irrigation abstractions, hence the provision of water storage will become increasingly necessary for irrigation.

The subject of water provision for irrigation is here dealt with only briefly. Bulletin 202, *Water for Irrigation** should be consulted for further information concerning the estimation of flow and the design and construction of storage.

Licences

Except in an area exempt from control, as mentioned previously, a licence must be obtained from the appropriate River Authority before abstracting water from any lake or watercourse or underground source of supply for the purpose of spray irrigation, or before increasing an existing abstraction. If, however, the water is then stored in an offstream reservoir, the subsequent abstraction from the reservoir is not subject to control by the River Authority, provided the reservoir is not a 'source of supply' for the purposes of the Water Resources Act 1963, but a licence would have been needed to fill the reservoir.

A reservoir is a 'source of supply', for example, if it is an excavation into underground strata, where the level of water in the excavation depends wholly or mainly on water entering it from those strata. It may also be a source of supply if water discharges from it to a watercourse, lake, estuary or the like, or if it is filled by water from land drains even when these are wholly used under the abstractor's own land. A licence must also be obtained before commencing to construct or alter any works for impounding or diverting the flow in a watercourse, and before commencing works for the abstraction of underground water. These works in relation to underground water include not only the construction or extension of wells and boreholes, but also the installation or modification or any machinery or apparatus for the abstraction of underground water.

A fee is payable to the River Authority on the granting of each licence and annually for licences to abstract water. In addition, charges are levied by the Authority based on the quantity of water licensed for abstraction.

---

* Bulletin 202. *Water for Irrigation*. (H.M.S.O.) Reprinting.

Because there are likely to be wet seasons when the quantity of water licensed is not used, farmers and growers may prefer to be charged on a special two-part tariff available to spray irrigators. Under this there is a basic charge for the licensed quantity of water, with supplementary charges based on the quantity of water taken from time to time.

One of the effects of the Water Resources Act, 1963 was to abrogate common law rights in relation to the abstraction of water. In their place the Act confers on the licensee a 'protected right' to the quantity of water authorized for abstraction. Thus River Authorities have a statutory duty not to issue additional licences, or to take other action which would prejudice the availability of sufficient water to meet the requirements of licences for the time being in force.

There is an important limitation to the degree of protection afforded to spray irrigators by a licence. In the event of exceptional shortage of rain or other emergency, a River Authority may serve notices reducing for a specified period the quantity of water that may be abstracted for spray irrigation. To avoid the dangers of water being cut off when most needed, farmers are advised to construct their own reservoirs, so that water can be taken when flows are high and used for irrigation in the summer months. This limitation in emergency does not apply, however, to abstractions of water for use in glasshouses, or to water used in the application of nutrients, pesticides and the like.

Surface Water

Surface water is present on most farms in one form or another. It may be a major river taking the drainage of thousands of square miles or, at the other extreme, just the underdrainage from the farm itself. On some farms there will be existing storage, either natural or artificial, in the form of lakes or ponds. Whilst only a small proportion of farms are in the fortunate position of having enough water available for irrigation by direct abstraction throughout the summer, there are many where the total annual flow is sufficient if winter storage is feasible, and provided a licence is granted. An estimate of the surface water available must be made to see if the irrigation requirements can be fully met, bearing in mind any limitations imposed by the licence.

Estimation of Water Available

If the source is a river or large stream apparently with plenty of water, it is advisable to consult the appropriate River Authority as to the amount of water actually available. For sources such as small streams, where little is known about the water available, the flow must be estimated as accurately as possible. Frequent flow gaugings are the best guide and the next best is an estimate based on the stream catchment area and likely runoff. An estimation of the flow available from all possible sources should be made at the earliest opportunity so that subsequent decisions are based on the best possible information.

Flows in small watercourses can most conveniently be measured with the aid of a V-notch or rectangular weir (Fig. 5). A V-notch weir is the best device for measuring small flows. A 6-in. V-notch is suitable for flows up to 10,000 gal per hr, and a 12-in. V-notch will measure flows up to 50,000 gal per hr. For higher flows a rectangular weir is preferable. The notch or weir can be made of metal, timber or other material of adequate strength, but for accurate gauging the crest should be sharp; this can be achieved by using thin material or bevelling the edge of thick material on the downstream side. The weir or notch should be positioned in a straight and uniform length of channel, set at right angles to the direction of flow and extended well into the banks and stream bed for watertightness and stability.

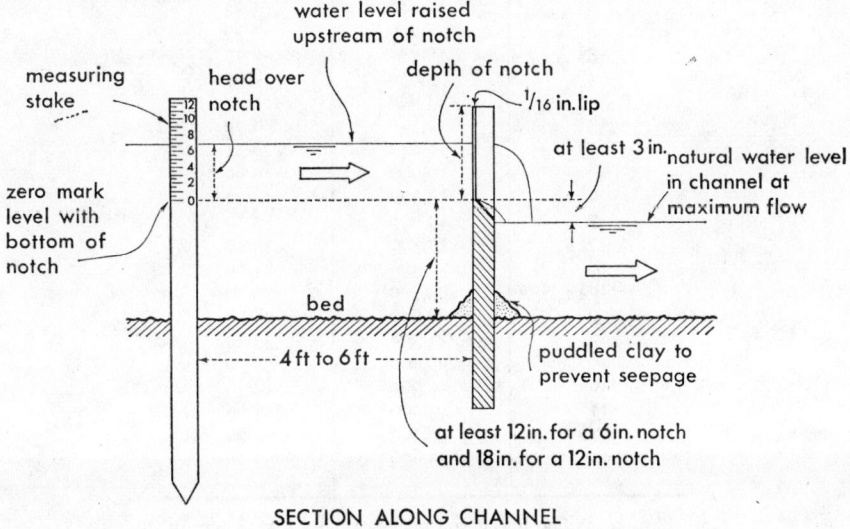

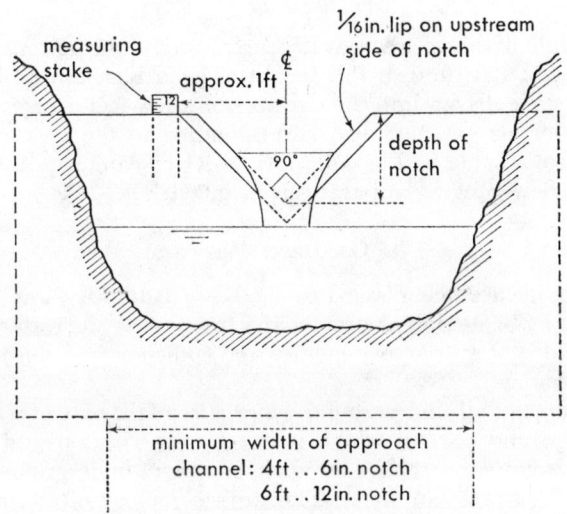

Fig. 5. Details of installation to give reasonably accurate gaugings

Table 6

Waterflow in relation to measured head
90° V-notch flow gauge head and flow readings

Head in in.	Gal per hr	Acre-in. per day
½	25	0·03
1	115	0·12
1½	310	0·33
2	650	0·69
2½	1,140	1·21
3	1,790	1·90
3½	2,620	2·77
4	3,650	3·87
4½	4,880	5·17
5	6,350	6·72
5½	8,040	8·51
6	9,980	10·60
6½	12,200	12·90
7	14,600	15·50
7½	17,300	18·40
8	20,400	21·60
8½	23,700	25·20
9	27,300	28·90
9½	31,200	33·10
10	35,400	37·50
10½	40,000	42·40
11	45,000	47·60
11½	50,100	53·20
12	55,600	59·00

Flow = $55,600\,h^{2\cdot48}$ gal per hr = $58\cdot9\,h^{2\cdot48}$ acre-in. per day where h = head over notch in ft.

The height of the upstream water surface above the bottom of the V-notch determines the flow and, if this height of head is measured by means of a graduated stake driven into the stream bed a few feet upstream, the flow in gal per hr or acre-in. per day, corresponding to the measured head, can be found from Table 6. The appropriate River Authority must be notified before any flow-gauging apparatus is installed.

Ground Water

Subject to a licence being issued by the River Authority, any type of well or bore-hole may be used to abstract water from below the water table, providing the yield is of satisfactory quality and sufficient in quantity to meet the irrigation demand.

The source of all underground water is rainfall, part of which is used in transpiration and part of which is lost by evaporation and run off. The remainder eventually reaches the main supply of underground water in the water-bearing strata (aquifers). The surface geology of an area determines the extent to which rainfall will run off as streams or seep into the soil. The ability of aquifers to yield ground water into a well, either hand-dug or bored, depends upon the nature of the spaces between the constituent particles, as, for example, the dimensions of pores and fissures and the degree of intercommunication. Water is stored within rocks in such cavities.

The geological pattern of this country is extremely complex, the rocks ranging from wholly impermeable to freely-yielding pervious water-bearing strata. As a generalization, the most important sources of underground water in England and Wales are south-east of a line from Tynemouth to Torquay, where younger geological strata predominate, and in an area of the Midlands, Cheshire and South Lancashire. The principal water-bearing strata are the Chalk and Triassic sandstones, together with the Lower Greensand and the Oolites. Locally supplies occur in drift deposits such as gravels, usually of shallow origin. Fig. 6 shows the extent of the principal water-bearing strata in England and Wales and their underground extension.

Geological advice on the prospects of obtaining ground water from a bore-hole or well, and its chemical quality, can be obtained from The Secretary, Water Resources Board, Reading Bridge House, Reading, Berks. From an evaluation of the geology and on the advice of a competent well-sinker, the details of a scheme for abstracting ground water can be determined. The depth of a bore-hole will depend upon the depth and thickness of the water-bearing strata and may lie between 20 and 400 ft, although deeper bore-holes are occasionally required. The diameter of bore-holes for irrigation purposes will be between 6 and 18 in., but the minimum diameter is usually governed by the size of pump required. Where water-bearing strata are fine-grained and unconsolidated, tending to 'run' when abstraction takes place, slotted lining tubes, sand screens or gravel packs are required. The length of lining tubes depends upon all these factors and on the need to prevent ingress of surface pollution, or other waters of adverse chemical quality. Linings may be constructed of cast iron, steel, aluminium, plastic, concrete, asbestos cement or brickwork, depending upon the circumstances. Shallow sources of underground water can sometimes be utilized by means of a series of tube wells ('well points') a short distance apart and connected to a common pump.

The temperature of underground water from deep sources is fairly constant, varying within a few degrees of 11°C, whereas that from shallow sources has a wider range, though the variation is less than that of air temperature.

As ground water percolates through water-bearing strata, the chemical quality of the water tends to change. In some cases, the resulting quality may be unsuitable for particular crops, as, for example, when high concentrations of sodium and chloride ions are present. Contamination of underground water near estuaries and coasts by saline infiltration may preclude its use for irrigation.

Yield

Although the abstraction licence will state the quantity of water authorized to be abstracted from the well or bore-hole during the period or periods specified, a pumping test must be made to determine whether the well or bore-hole has sufficient potential to meet the demand. The yield of dug shafts and bore-holes depends upon so many geological and constructional factors that it is not possible to generalize. Accurate measurements of water levels and the yield of the boring should be recorded. The test pumping

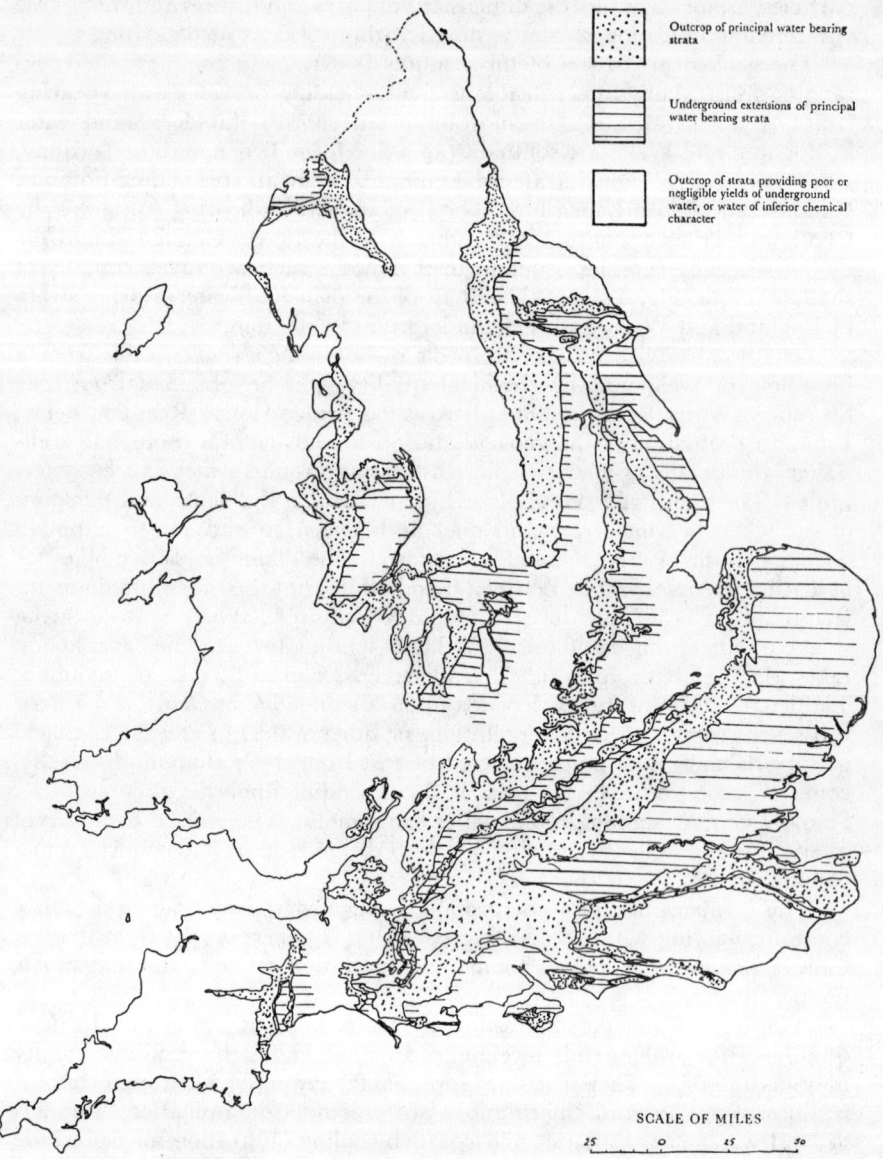

Fig. 6. Principal water-bearing strata in England and Wales and their underground extensions

rate must exceed the rate at which water will be required for irrigation and the test must be continued long enough, at least 24 hr, to indicate whether such a rate can be maintained. The information obtained during a pumping test will be useful in determining the final details of the permanent equipment to be installed.

During pumping, the water level is lowered in the water-bearing strata around the well. The effect will be greater as the abstraction rate increases, and there may be interference with the supply to neighbouring installations.

Public Supply

Information on the possibility of using public supply water for irrigation can be obtained from the local Water Undertaking, who can give details of the quantity, pressure and cost of any supply which may be available and advise whether supplies can be assured in dry periods. The capacity of rural water mains is usually insufficient for a direct supply to a farm without upsetting the functioning of the system elsewhere. Unless the requirements are very small, it will usually be necessary to provide storage equal to a day's requirement at peak demand so that water can be stored overnight, thus reducing the effect on mains pressure during the day. The cost of the water is likely to be too high for the lower value crops such as grass and sugar beet, but may well be economic for high-value crops such as early potatoes and market-garden crops, for which the acreages and water requirements tend to be much smaller.

STORAGE OF WATER

Because irrigation water demands occur at times when water is least available and are concentrated into a few months or weeks of the year, storage will be necessary in connection with most irrigation projects.

The subject of water storage is dealt with in detail in Bulletin 202,* *Water for Irrigation*, to which reference should be made for further information. The Reservoirs (Safety Provisions) Act, 1930 requires that a reservoir designed to hold, or capable of holding, more than 5 million gallons of water above the natural level of any part of the land adjoining the reservoir must be designed, constructed under the supervision of, and inspected from time to time by, a qualified civil engineer or one of the panels constituted under the Act. Local conditions vary to such an extent that a reservoir project of any significance requires the services of a qualified civil engineer who has experience in this field.

For small schemes supplied from the public mains, over-night storage of a few thousand gallons in a tank may be all that is needed whilst, at the other extreme, an earth reservoir of several or even tens of millions of gallons may be required. Where a licence is restricted to winter abstraction from a stream or river, then the whole season's requirements must be stored, but there is not necessarily any direct relationship between the irrigation demand and the storage capacity needed. The latter may be large or small to meet the same irrigation demand, depending upon the reliable flow into the reservoir.

Sites for earth reservoirs exist on many farms. The best arrangement is an offstream reservoir which is filled by pumping or, if the site levels permit, by gravity. The other type of reservoir is the familiar impounding reservoir which is constructed by building a dam across a watercourse. These are much more difficult to site and construct than off-stream reservoirs. The latter arrangement offers much more opportunity to find a site close to the irrigation area, and where other important site characteristics such as impermeability are favourable. The water-retaining characteristics of an earth reservoir site are a major consideration as the cost of artificially waterproofing, by lining or other means, greatly increases the cost of storage.

---

* Bulletin 202, *Water for Irrigation*. (H.M.S.O.) Reprinting.

A seepage reservoir, i.e., an excavation into the water table in a shallow permeable deposit, combines the functions of a well and storage. An existing gravel pit may be a useful ready-made source for irrigation water supply but, where such ready-made facilities do not exist, it may be quite sufficient to sink a well or wells to abstract water which is already stored in the natural reservoir of the gravel itself.

WATER QUALITY

The chemical and biological quality of water may affect its suitability for irrigation. Some pollution can be tolerated, but it is difficult to determine the degree of pollution and to define a permissible amount. During planning, seriously contaminated water sources will have been avoided but during prolonged droughts sources which are normally acceptable can become seriously contaminated. The degree of contamination should be checked during long dry periods to make sure that it has not become excessive and that the water is safe for use on the crops being irrigated. Further assistance may be obtained from the Agriculture Development and Advisory Service.

CHEMICAL QUALITY OF IRRIGATION WATER

No water used by crops is pure, not even rainwater. Most materials present in water sources are quite harmless to crops though they vary in their concentration. Some waters are hard, others soft and some contain nutrients beneficial to crop growth. A good source of water for irrigation is free from harmful organisms and does not contain dissolved or suspended matter at a concentration detrimental to crop growth or to the physical condition of the soil.

Many water sources carry solid particles in suspension such as sand, clay, humus or iron. An undue amount of suspended solids in the water will necessitate frequent cleaning of filters and nozzles. If the solids are abrasive they will cause excessive wear to the moving parts of the equipment, such as pumps and oscillators. The suspended particles may form an unattractive deposit on some horticultural crops such as green vegetables or flowers.

Chemical salts dissolved in water supplies vary widely both in their nature and concentration. Sulphates, carbonates and bicarbonates of calcium and magnesium are present in nearly every ground water supply. While the total concentration may vary from a very small quantity up to 0·2 per cent, these salts are not usually harmful to crop growth. They may, however, increase the alkalinity of soils, and 5 in. of water containing 0·05 per cent calcium carbonate will add the equivalent of 5 cwt per acre of chalk or lime. Some waters contain moderate amounts of iron salts which, while not causing damage to crops, may result in unacceptable coating of leaves or fruit and cause blockages in nozzles. The water may be treated to reduce the iron content by aeration followed by settling or filtration, but the capital outlay may be substantial.

Sources contaminated with sewage or industrial effluent may contain salts of certain metals, such as chromium, zinc or copper, or other chemicals, such as borates, which may have a deleterious effect on crops. Irrigating with water containing detergents can also adversely affect crop growth but, with a few exceptions, natural river water in this country has not yet been found to contain sufficient detergents to have any significant effect when used for ordinary irrigation.

Any chemical effect on non-ferrous pipe systems is comparatively rare. If the water is suitable for use on crops, it will probably not affect asbestos cement, plastics or aluminium pipes. Water containing traces of copper, however, will have a corrosive effect on aluminium equipment. Organic slurries, if passed through irrigation systems, may have a corrosive action. Very soft waters tend to be acid in character and corrosive to steel pipes and fittings. Galvanizing is some protection, but will not withstand the effect of strongly acid water. This aspect is less important, now that plastic pipes and equipment are available. River water containing detergents has been known to wash out the lubricants from irrigation oscillators and pumps.

Some chemical substances, e.g., sodium chloride (common salt), if present in sufficient amounts, may have a damaging effect on crop growth, soils and irrigation equipment. Brackish or saline water should not be used for irrigation, though water containing less than 0·05 per cent sodium chloride is usually regarded as being safe for all purposes (0·05 per cent would supply 1 cwt/acre salt for each acre-in. of water).

Water containing salt may damage the physical condition of the soil and local advice should be sought on this point. The sensitivity of crops to salt varies widely and depends a good deal on the stage of growth of the crop. Crops that are particularly sensitive include apple, pear, plum, strawberry, red currants, lettuce, peas and beans, cocksfoot, potatoes and most clovers. For more specific advice on the use of saline waters, local Advisory Services should be consulted.

It is essential to ensure that any sample of water taken for analysis is representative of the supply to be used for irrigation; it must be remembered that any contamination is likely to be greater during summer when the flow is least.

Biological Quality of Irrigation Water

The term biological quality here refers to the aspect of water supplies used for normal irrigation. It does not relate to the practice popularly known as 'organic irrigation' which is a method of farm waste disposal unrelated to soil moisture requirements and therefore outside the scope of this bulletin.

POLLUTION PREVENTION LEGISLATION

Under the Rivers (Prevention of Pollution) Acts 1951 and 1961 it is unlawful to discharge any poisonous, noxious or polluting matter, trade or sewage effluent into a stream without the consent of the River Authority. For the purposes of the Acts a 'stream' is defined as 'any river, stream, watercourse or inland water, but not a lake or pond which does not discharge into a stream'. The Water Resources Act 1963, Sections 72–75 makes it an offence to discharge into any underground strata any trade or sewage effluent, or any other pollution matter without the consent of the River Authority.

CROPS

Even with these extensive powers much remains to be done to bring surface waters up to a satisfactory standard and pollution risks can arise, with consequent health hazards. Crops normally eaten raw such as market-garden, salad and fruit crops constitute the greatest risk to human health if irrigated with polluted water; sources of water used for irrigating these crops should first be subjected to a biological examination to determine their suitability. Water from grossly polluted sources should not be used; others may require the installation of chlorinating or similar equipment and the washing of the produce with chlorinated water. To minimize the risk, irrigation should cease well before harvesting.

GRASSLAND

It is important to remember that some animal diseases can be spread by grazing. It is essential to check the source of water used for the irrigation of grassland and if doubts arise a veterinary opinion should be sought.

The present state of knowledge suggests that the danger from pollution to livestock from applying irrigation water is remote. The nature and degree of pollution is obviously relevant, but in the exceptional circumstances of the water being grossly polluted and likely to be carrying *Salmonella* organisms, it should not be used on pasture or fodder crops within six months prior to their being grazed by either cattle or sheep. Such conditions might arise from the remote pollution of a stream or from accidental contamination of private storage by effluent from known carriers on the same farm. Brucellosis and Johnes infections can be similarly transmitted and the disposal of infected material should be the subject of veterinary advice.

Estimation of Potential Yield of a Water Source

The estimation of the amount of water which may be available from a source is referred to in previous pages and further information is given in Bulletin 202. A number of factors influence the seasonal potential yield of a well or flow in a stream, but a limiting factor is the *excess* winter rain on the catchment area of an aquifer or stream.

Stream-flow and natural underground water storage are generally increased or replenished by the rainfall during the winter half-year which becomes available when the soil has returned to capacity. This 'excess' winter rain either runs off the surface, is removed via the drainage system, or moves downward through deep seepage to the water-table.

This excess winter rain can be calculated and a complete analysis is contained in Technical Bulletin 24,* *The Significance of Winter Rainfall over Farmland in England and Wales*. The analysis therein was computed for non-irrigated land.

Put in its simplest form 'excess' winter rain is the rainfall less the evaporation during the period (usually small) and less the soil moisture deficit at the end of the summer. In the case of irrigated land this end-of-summer deficit is less than that over non-irrigated land and the 'excess' winter rain is correspondingly greater. This means that more water would be available for

---

* Technical Bulletin 24, *The Significance of Winter Rainfall over Farmland in England and Wales*, price 45p (by post 50½p), obtainable from H.M.S.O. or through any bookseller.

storage in the locality but that throughput of water would cause more leaching.

Leaching Factor

This 'excess' rainfall is a good representation of the average leaching factor and exercises a strong control on liming policy. Its year-to-year value is also a great help in determining fertilizer requirements.

The map (Fig. 7) shows the average value of this leaching 'excess' rain for England and Wales; Scotland has not been included because of the rapid increase with height above sea level. The amount of 'excess' rain varies from an average of 3–4 in. in the driest parts of eastern England to 20 in. and more in the hills uplands and mountains of the west and north.

The variations about these averages from year to year are as great as 50 per cent; in the driest areas about one year in 20 is so dry that there is no complete replenishment of the summer soil moisture deficit and there is little or no drainage or leaching.

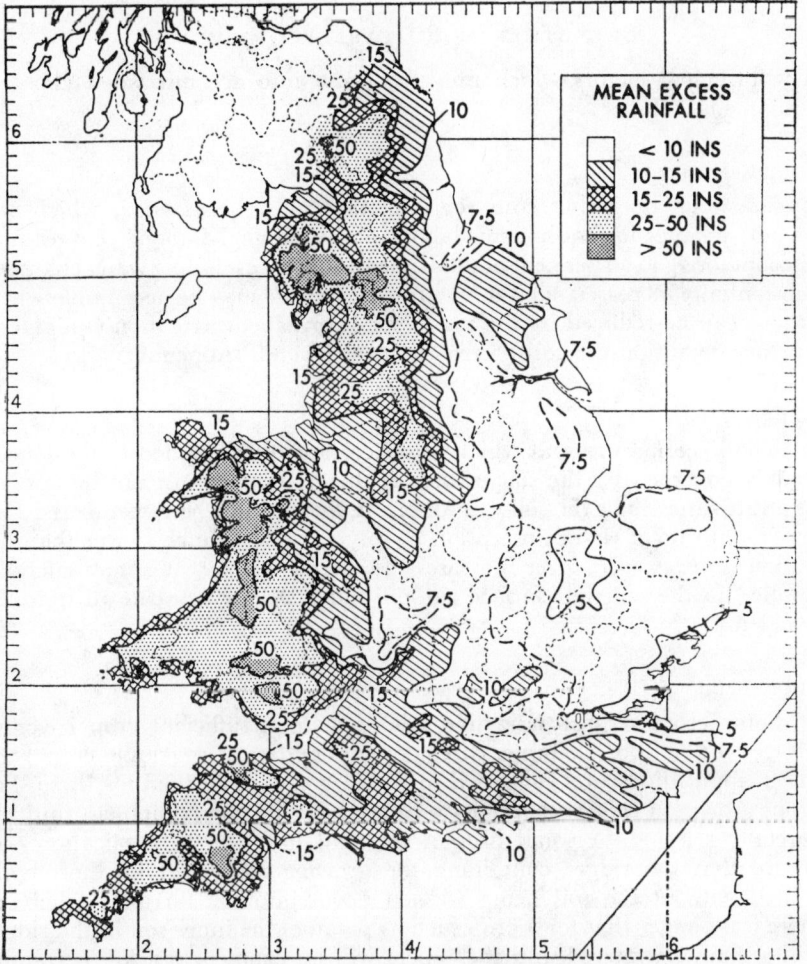

Fig. 7. Mean excess rainfall over England and Wales

Equipment (Outdoor)

DISTRIBUTION EQUIPMENT—TYPES AND SELECTION

THREE main types of distribution equipment are used in Britain: spraylines, used mainly in horticulture; medium to low pressure rotary sprinklers having nozzle diameters less than $\frac{1}{2}$ in., which cater for the bulk of agricultural needs; and high pressure rotary rain-guns and sprinklers having a nozzle over $\frac{1}{2}$ in. in diameter, which can water very large areas but which have not been widely used in Britain, apart from liquid manure distribution. Both low and high pressure systems may be mounted on self-propelled equipment.

SELECTION OF SUITABLE EQUIPMENT

The principal factors which must be taken into account when choosing equipment include:

EQUIPMENT COSTS

The total operating costs for irrigation must be assessed, which will include charges for equipment, labour and pumping. Labour charges may account for 25 to 30 per cent of total irrigation cost (excluding water charges); substantially increased capital investment per acre may be acceptable where labour can be reduced by the use of semi-permanent irrigation equipment. Further details on the cost of irrigation are given in Appendix V.

SOIL TEXTURE

The maximum rate at which water can safely be applied to the land is mainly governed by the soil texture, Table 7. Water must not be applied at a rate faster than the soil can absorb it, and it must be remembered that the initial rate of water absorption by any dry soil is much slower than the normal rate. A crop cover increases the rate at which water may safely be applied to the soil, but surface run-off and puddling must at all times be avoided.

CROP TYPE

Some crops require irrigation before there is sufficient crop cover to protect the soil, for example, early potatoes and most vegetable crops; the lower application rates suggested in Table 7 should be used for these. For crops grown on ridges, especially on unstable soils, the equipment must be carefully selected. A small droplet size and low rate of application will ensure that the ridges containing the growing roots will be thoroughly wetted without the soil being washed down into the furrows; otherwise there is a danger that for a crop such as potatoes the inner soil in the ridges will remain unwatered and the tubers will be exposed, quickly becoming green and unsaleable.

Table 7
Suggested application rates using commercial spraylines and sprinklers

Soil texture	Intake rate		Application rate	
	in. per hr	mm per hr	in. per hr	mm per hr
Coarse sands	0·5–1·0	12·7–25·4	0·5–0·8	12·7–20·4
Fine sands, loamy sands . .	0·5–0·8	12·7–20·4	0·5–0·75	12·7–19·1
Coarse sandy loams and fine sandy loams . . .	0·3–0·75	7·6–19·1	0·3–0·6	7·6–15·2
Very fine sand loams, sandy clay loams and silt loams . .	0·25–0·4	6·4–10·2	0·2–0·35	5·1–8·9
Clay loams and silt clay loams .	0·2–0·3	5·1–7·6	0·15–0·25	3·8–6·4
Sandy clays, silty clays and clay .	0·05–0·15	1·3–3·8	0·15 or less	3·8 or less

Notes:
1. The suggested application rates apply to bare soil: with crop cover, rates may be increased by up to 20 per cent.
2. The suggested application rates are for level ground and slopes up to 1 in 20: for slopes from 1 in 20 to 1 in 10 application rates should be reduced by 50 per cent.
3. Because small drops damage soil structure less than large ones it is possible to increase the application rate slightly if the equipment produces drops which are considerably smaller than those from spraylines and sprinklers in general use.

FIELD SHAPE AND TOPOGRAPHY

It is easiest to lay out and to move sprinkler and sprayline on rectangular parcels of land as the lengths of the lines remain unaltered when they are moved. On areas of irregular shapes rain-guns are easier to manage than sprinkler lines but have the severe limitations of high application rate and large droplet size.

As far as possible sprinkler laterals and spraylines should be laid across the slope of the land to reduce pressure differences and consequent differences in nozzle output. Differences of pressure (due to levels) of more than 10 per cent are considered unsatisfactory. Sprinkler laterals should if possible be laid at right angles to the prevailing wind direction to improve distribution, but this is often rendered difficult by other factors such as slope, which has just been mentioned. Pressure regulators can be fitted to individual sprinkler standpipes to reduce excessive pressures.

For rectangular plots, spraylines and 'rain fans' have some advantage in that they can be made to water a rectangular area, thus avoiding interference with crops on neighbouring land.

It may be worth modifying field boundaries to suit irrigation systems, rather than the reverse.

POWER AVAILABILITY

Large nozzle rain-guns operate at higher pressures than sprinkler or sprayline systems and thus have a higher power requirement.

For pumping, electric power is the most convenient and easy to maintain but diesel power is often cheaper and more mobile. Provision of a suitable electricity supply for pumping can sometimes be expensive. Farm tractors are widely used for pumping but their engines are not rated for continuous running and, if long periods of pumping are contemplated, not more than 60 per cent of the maximum power available should be used.

LABOUR

Use of labour is one of the major factors influencing the design of irrigation schemes including the rate of application of water, pipe sizes and capital investment. High application rates from rain-guns and large sprinklers means that virtually full-time attention is required. The recent trend towards lower rates allows less frequent attention and can fit in with meal times or milking times. The consequent savings in labour costs and the reduction or elimination of the arduous and unpopular task of moving pipes through wet crops must be set against the increased capital cost of extra equipment required, including extra piping.

Advances have been made with sprinklers mounted on booms that are self-propelled over the land. The main types available may either water a 'swath' as the equipment moves continuously across the field or water a large circular area from a rotating boom which is moved at intervals across the field.

WATER SUPPLY

A water supply should be free from suspended sand, algae, bacterial slime, or other impurities as these could lead to operational difficulties in the distribution system by causing blockages in fittings and nozzles. The installation of a filter may be required or treatment to the supply at source may be necessary.

Types of Distribution Equipment

SPRAYLINES

Spraylines may be made of aluminium, galvanized steel or plastic tubing of $\frac{1}{2}$ to 2 in. internal diameter. Nozzles are inserted along these pipes of a size and shape to give the required range of output and spray pattern. The sprayline irrigates a rectangular area which varies between a 30 and 50 ft width depending on the operating pressure. The distance between spraylines is related to the length of the jet issuing from the nozzles; to ensure an even distribution of water over the whole area there must be a slight overlap of the wetted areas from each line. The output of the nozzles along a sprayline decreases as the pressure loss along the line increases; the magnitude of this pressure loss depends upon the nozzle spacing and output, and the length and diameter of the sprayline. In practice it is usual to limit the pressure loss along a line to 25 per cent of the average operating pressure. For example if the average line pressure is 40 lb/sq. in., 45 lb/sq. in. and 35 lb/sq. in. would be acceptable at the two ends. Nozzles are generally spaced 2 ft apart. The line can be continuously oscillated by water motors which use water pressure to operate single or double acting pistons, or by a water filled counterweight system. In an attempt to dispense with oscillators, spraylines have been produced with a series of jets at different angles. The nozzles must be closely spaced to achieve acceptable distribution patterns and this may lead to high-application rates or nozzle blockage if low-output nozzles are fitted.

Spraylines consist of sections which usually have quick couplings to facilitate dismantling for moving. They are mounted on stands, usually 2 ft high to raise them above the crops. For crops of greater height such as stick beans, tall stands are driven into the ground and left in position for the whole season.

Sprayline irrigation, outside glasshouses, is restricted largely to salad and

EQUIPMENT (OUTDOOR)

flower crops and very small scale production of vegetable crops. Outside crops on anything approaching field scale are generally watered by sprinkler.

SPRINKLERS

Various types of sprinkler heads are available having one or two nozzles from which the water issues. The water from the sprinkler is spread more evenly over the wetted circular area by the breaker; a spring-loaded weighted arm actuated by the water jet which both rotates the sprinkler head and breaks up the jet.

A typical line of medium pressure sprinklers employs sprinkler heads that water a circle of about 45 ft radius. If such sprinklers are spaced at 35 ft intervals along the lateral line and the line is moved 60 ft at a time, the area covered by a line 700 ft in length (20 sprinklers at 35 ft spacing) is 1 acre. If each sprinkler delivers about 6 gal per min the rate of application is about $\frac{1}{3}$ in. per hr. The present tendency is to use sprinklers giving lower application rates over a greater wetted diameter set out on a square pattern at up to 60 × 60 ft spacing. Precipitation rates obtained with different sprinkler spacings and outputs are given in Table 8.

TABLE 8

Precipitation rates in./hr with varying sprinkler spacings and output

Sprinkler output in gal per min	2	3	4	5	6	7	8	9	10	11	12	13	14
Spacing feet													
20 × 20	0·58	0·86	1·15	1·46	1·74	2·04	2·33						
20 × 30	0·44	0·58	0·77	0·96	1·15	1·36	1·55	1·74	1·93	2·12	2·31	2·52	
20 × 40	0·29	0·43	0·57	0·72	0·86	1·02	1·15	1·31	1·45	1·60	1·74	1·88	2·04
20 × 50	0·24	0·36	0·47	0·38	0·70	0·82	0·92	1·04	1·16	1·27	1·40	1·51	1·64
20 × 60	0·19	0·29	0·38	0·48	0·58	0·67	0·77	0·86	0·97	1·06	1·16	1·26	1·36
30 × 30	0·25	0·38	0·52	0·65	0·77	0·90	1·03	1·16	1·28	1·42	1·55	1·67	1·80
30 × 40	0·19	0·29	0·38	0·48	0·58	0·67	0·77	0·86	0·97	1·06	1·16	1·26	1·36
30 × 50	0·16	0·23	0·31	0·38	0·46	0·54	0·63	0·70	0·77	0·85	0·92	1·01	1·08
30 × 60	0·13	0·19	0·25	0·32	0·38	0·45	0·51	0·38	0·64	0·71	0·77	0·83	0·90
35 × 30	0·22	0·34	0·44	0·52	0·68	0·76	0·86	1·00	1·10	1·20	1·34	1·44	1·54
35 × 40	0·17	0·25	0·33	0·41	0·50	0·58	0·67	0·77	0·82	0·91	1·01	1·10	1·15
35 × 60	0·11	0·17	0·22	0·26	0·34	0·38	0·43	0·50	0·55	0·60	0·67	0·72	0·77
40 × 40	0·14	0·22	0·29	0·36	0·43	0·50	0·58	0·65	0·72	0·79	0·86	0·94	1·01
40 × 50	0·12	0·17	0·23	0·29	0·35	0·41	0·47	0·52	0·58	0·64	0·70	0·76	0·82
40 × 60		0·14	0·19	0·24	0·29	0·34	0·38	0·43	0·48	0·53	0·58	0·62	0·67
40 × 80		0·11	0·14	0·18	0·22	0·25	0·29	0·32	0·36	0·40	0·43	0·47	0·50
50 × 50		0·14	0·18	0·23	0·28	0·32	0·37	0·42	0·47	0·51	0·55	0·60	0·64
50 × 60		0·12	0·16	0·19	0·23	0·28	0·31	0·35	0·38	0·42	0·47	0·51	0·56
60 × 60			0·13	0·16	0·19	0·23	0·25	0·29	0·32	0·35	0·38	0·42	0·46
60 × 80				0·12	0·14	0·17	0·19	0·22	0·24	0·26	0·29	0·32	0·34

The presence of trees makes it difficult to achieve an even distribution of irrigation water in orchards. Sprinklers with single or double low-angle nozzles may be used but this reduces the length of throw. An alternative is to irrigate over the trees using standard sprinklers on tall risers. Low-angle sprinklers may also be used in very exposed situations to reduce the effect of wind on water distribution.

Sprinklers and laterals should be arranged at distances to give an overlap which ensures a reasonably even distribution of water, and spacings of 60 per cent of the wetted diameter, both between the sprinklers and the lateral pipes, are generally satisfactory. The number of sprinklers on any one line must be limited so that the drop in water pressure due to friction does not exceed 25 per cent of the average operating pressure, otherwise the area at the end of the line will be under-watered.

Table 9 can be used to determine the maximum number of sprinklers permissible on a lateral to keep friction loss to 25 per cent of the average operating pressure.

Table 9

Maximum number of sprinklers on sprinkler line

Pipe diameter 2 in.

Sprinkler spacing	20 ft			30 ft			35 ft			40 ft		
Av. operating pressure p.s.i.	25	35	45	25	35	45	25	35	45	25	35	45
4 g.p.m.	14	15	17	12	13	15	11	12	14	11	12	13
6 g.p.m.	10	11	13	9	10	11	9	9	11	8	9	10
8 g.p.m.	9	9	11	8	8	9	7	8	9	7	7	8

Pipe diameter 3 in.

Sprinkler spacing	20 ft			30 ft			35 ft			40 ft		
Av. operating pressure p.s.i.	25	35	45	25	35	45	25	35	45	25	35	45
4 g.p.m.	29	32	35	25	27	30	23	25	29	22	24	28
6 g.p.m.	23	24	28	20	21	24	19	20	23	18	19	21
8 h.p.m.	19	20	23	17	18	20	15	17	19	15	16	18

Pipe diameter 4 in.

Sprinkler spacing	20 ft			30 ft			35 ft			40 ft		
Av. operating pressure p.s.i.	25	35	45	25	35	45	25	35	45	25	35	45
4 g.p.m.	49	52	60	42	45	52	39	42	49	38	40	46
6 g.p.m.	38	40	47	33	35	40	31	33	38	29	31	36
8 g.p.m.	31	33	39	27	29	33	26	27	31	24	26	30

EQUIPMENT (OUTDOOR)

RAIN-GUNS AND LARGE SPRINKLERS

Rain-guns and large sprinklers have not been commonly used for clear water irrigation in Britain. The head is driven round by a spring loaded swinging arm or turbine which will also break up the jet. Compared with small sprinklers the output of water is much greater and for an adequate throw this type of equipment requires much higher supply pressures. The radius of the circle watered may be 60 to 180 ft and only large drops of water will reach the outside of the circle. These large drops, applied at rates which may approach 1 in. per hr, can cause considerable damage to the soil structure unless the equipment is used with care. Normally, rain-guns should be used only where there is full crop cover, as for example with grass or well-established sugar beet.

FIELD TRICKLE EQUIPMENT

Trickle equipment similar to that originally developed for glasshouses (see description on page 66) can be used in semi-permanent 'solid-set' layouts in the field, but is more expensive than sprinklers and spraylines and its use is therefore justifiable only for comparatively high-value crops.

Trickle equipment is also available for use in orchards where each tree is provided with one or two outlets. In one version the supply pipe consists of plastic hose, severed at each outlet point for insertion of the trickle device. This device comprises two concentric 6-in. plastic cylinders whose bore is identical with that of the supply hose; thus no restriction is introduced. The outside of the inner cylinder has a spiral groove communicating at one end with the bore; water flows along the groove under high friction and drips on to the soil at the outer end.

Drip-watering, in which the water emerges on to the soil through lengths of small bore plastic tubing inserted into the side of the main pipe (see page 68), may also be used for field crops.

Trickle systems have the important advantage that they leave most of the soil surface dry and therefore free from wastage of water by evaporation.

PIPES AND NOZZLES

WATER FLOW THROUGH PIPES

Pipes are usually described by reference to their internal (i.d.) or external (o.d.) diameters, depending on the material from which they are made. Irrigation pipes, being thin-walled, are generally known by their 'nominal size' and the difference between 'i.d.' and 'o.d.' is ignored. For strictly accurate work 'i.d.' is the important measurement.

As water passes through a pipe there is a pressure loss due to friction and the greater the rate of flow through a given pipe the greater will be the friction. It is a common fault to install a pipe too small for the proposed supply or one which will not allow for future expansion of the system or for a build-up of incrustation on pipe walls. With portable irrigation mains there is seldom any practical advantage in reducing pipe sizes as the main branches and potential flow decrease, because of the inconvenience of sorting out pipes of different sizes.

The following terms are commonly used in discussing pipe size and water flow.

HEAD OR PRESSURE HEAD

This is quoted either as a pressure, lb per sq. in. or as ft head, and a pressure of 1 lb per sq. in. is equal to a head of 2·31 ft of water. A slope may either increase or decrease the pressure head, depending on the direction of flow of the water. Thus the pressure head which results from an overhead water tank is the height of the water surface in the tank above the point at which the water issues from the pipe.

FRICTION PRESSURE DROP

Loss of pressure due to friction is normally quoted as 'loss of ft head per 100 ft of pipe run.' The relevant figure can be obtained from tables for any given rate of water flow, pipe size and material. Other parts of the pipe system such as bends, valves and fittings make their contribution to pressure loss, and a small allowance (usually 10 per cent of the pipe friction loss) should be made if there are very many in the system and especially where small diameter pipes are being used. Table 10, which should not be used for design purposes, gives a rough guide to the selection of suitable pipe diameter related to flow and distance.

Table 10

Guide to internal diameter of pipe (in.)

Gal per min	Length of pipe, ft			
	0–500	500–1,000	1,000–2 000	above 2,000
5	1	1¼		
10	1¼	1½		
25	2	2½		
50	2½	3	3	
100	3	3	4	
150	3	4	4	
200	4	4	5	6
250	4	4	5	6
300	4	5	6	6

Main Pipes

The function of the main pipes is to convey water from the source to the application equipment in the quantity and the pressure required for correct operation. Permanent main lines of metal or plastic materials are buried 18 in. deep when the pipes are of 2 in. diameter or less and 30 in. deep when the pipes are larger than 2 in. diameter. The alternative of above-ground light-weight mains may be cheaper and more easily moved, but they can be a nuisance to cultivation and transport equipment.

A 'ring main' is sometimes installed, the principal advantage being that water can flow from two directions to the take-off point, so that friction losses are smaller and smaller pipes may be used if desirable. However, a ring normally requires extra lengths of pipe and not many situations lend themselves to this layout.

The two most important factors in main-line design are the relative costs of each size of piping and the capital and operating costs of pumping water through the different sizes of piping. There is no fixed rule on how much to allow for the pressure loss due to friction. Figures between 15 and 30 per cent of the total head are common but may be often exceeded in practice. The power cost of friction loss should always be balanced against the cost of the equipment.

Since pressure loss due to friction depends on the distance the water has to travel, pressure loss will be greater where a sprinkler is operating at the far rather than the near end of a main. For this reason when two laterals are operating from one main they should start at opposite ends. Initially half the water will flow to the far end, and as the laterals are moved to the centre of the main the full quantity of water will have to travel half the distance (Fig. 8).

The following points are important when considering pipe sizes:

(1) For a given pipe diameter and quantity of water passing through it, loss of pressure is proportional to the length of the pipe;

(2) for a given size and type of pipe the pressure loss due to friction is approximately proportional to the square of the flow rate. Thus doubling the flow in a given pipe results in the pressure loss being increased about four times; and

(3) loss of pressure due to friction does not depend on the initial pressure of the water. The loss would be the same if the initial pressure was 10, 25, or 50 lb per sq. in.

Tables and graphs are available from which the friction loss appropriate to the type and size of pipe and the flow rate can be read off.

Hydrants and Standpipes

At intervals suited to the throw of the particular distribution equipment T connections with valves and couplings are provided for attaching the sprinkler laterals (Fig. 9). A sprinkler line may be connected directly to the take-off point, or connected through a length of high pressure flexible hose. The hose connections are of particular importance with the small-bore semi-permanently placed systems of laterals.

When the main is buried a riser is fitted so that the connection may be made at or above ground level. For ground level connection, the valve and coupling may be housed in a small chamber having a removable lid flush with the soil surface. This arrangement is known as a hydrant and to prevent damage, especially to and by cultivation equipment, each one should be clearly marked. Alternatively, there may be a riser and valve, perhaps 2 or 3 ft above ground level; this is known as a standpipe.

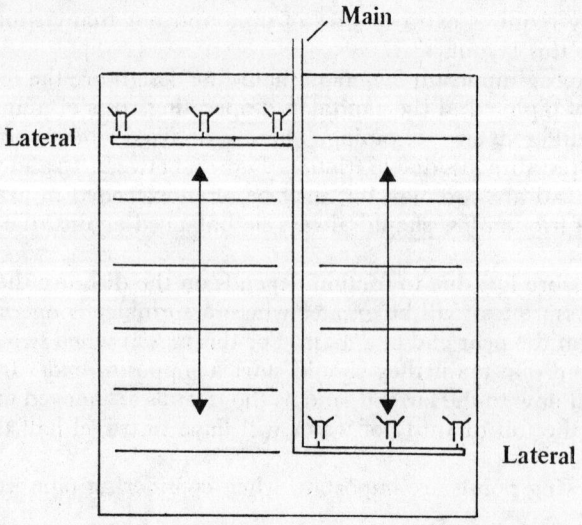

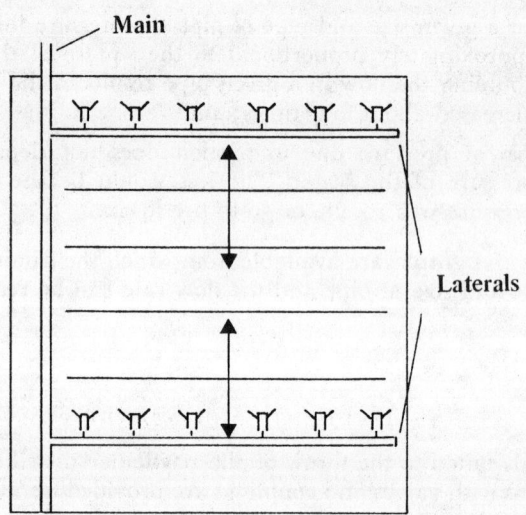

Fig. 8. Two suggestions for irrigation of rectangular areas using two laterals

Lateral Pipes

This term is usually applied to the portable lines which carry the water via take-off valves from the main pipe to each point where the water distributors are located. At the end of each line a plug or 'stop end' is inserted. There are other fittings which can be used with the portable equipment such as bends, tees and crosses.

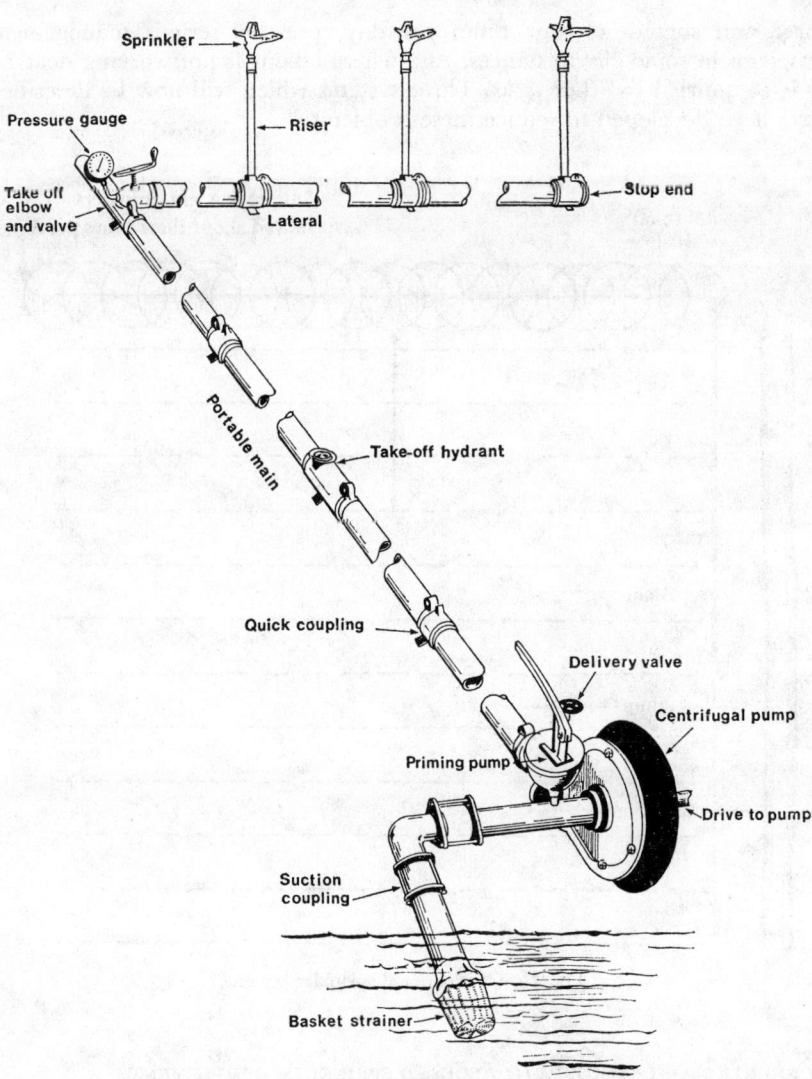

Fig. 9. A sprinkler layout showing principal components

There is a considerable range of choice in lateral pipe layout, the selection generally depending on a balance between labour requirement of the system and capital investment. The amount of capital invested can be considered in four stages:

CONVENTIONAL LAYOUTS

Here each lateral carries a full complement of sprinklers and is moved each time the sprinklers have applied the required quantity of water. This normally means three moves a day (Fig. 10). This system is very efficient in terms of equipment utilization and accordingly the capital cost is low. However the need to move the sprinkler lines at regular intervals, often three and sometimes four times per day, presents serious management problems in some circumstances, e.g., where labour is not working near by or is committed to other tasks. Three systems which will now be described have been developed to reduce these problems.

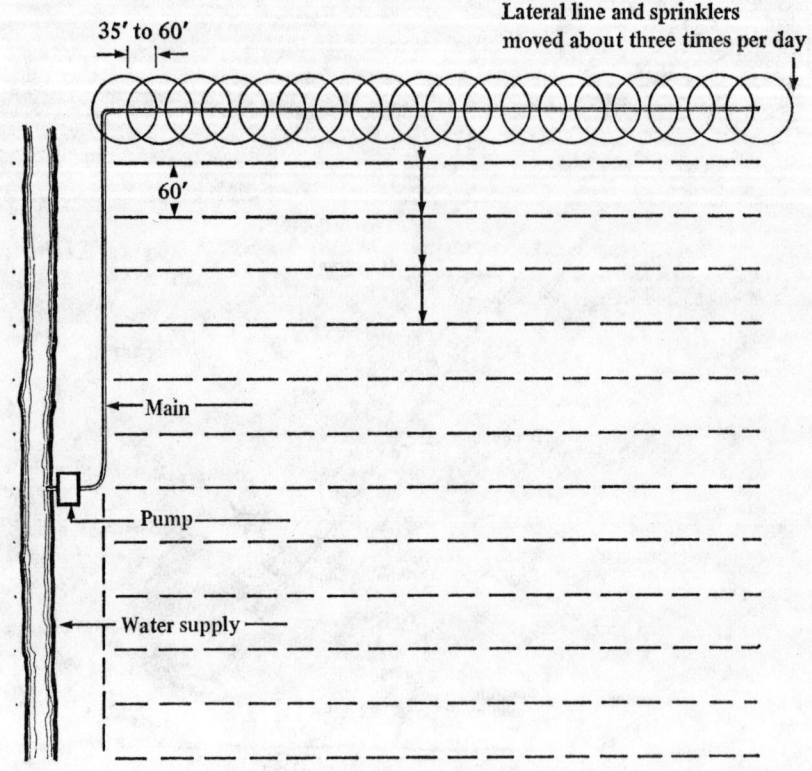

Fig. 10. Conventional sprinkler layout

CONVENTIONAL LAYOUT WITH REDUCED SPRINKLER COMPLEMENT

In this system the portable laterals are of the same type and are laid out in the same way as in the conventional system. However, each sprinkler standpipe connection on the laterals is fitted with a valve which automatically opens when the standpipe is inserted and closes when it is removed. Sprinklers, usually of the long-range low-precipitation-rate type, on standpipes are inserted in alternate standpipe connections along the laterals. The equipment is operated for sufficient time to apply the desired quantity of water, and then without shutting off the lateral each sprinkler and standpipe is moved to the alternate position (Fig. 11). This operation is normally done once per day and the laterals moved each morning, thus avoiding the need for more than very minor attention during the day. Although extra lateral piping is required than with the conventional system, the pipe diameter can often be reduced. Also, the longer irrigation period per day reduces the pumping rate which in turn can reduce the cost of this equipment.

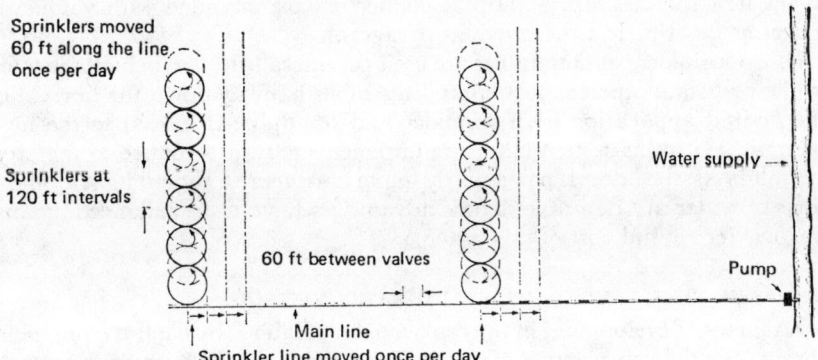

Fig. 11. Conventional layout with reduced sprinkler complement

SEMI-PERMANENT LAYOUTS (SOLID-SET)

In this system a series of small-bore laterals ($1\frac{1}{4}$ in. diameter) is placed in position for the duration of the crop. Each lateral is connected to the main which runs along the side of the field by a flexible hose fitted with a quick coupling (Fig. 12). This allows the hose to be moved out of the way of tractors turning on the headland during cultivation and spraying operations

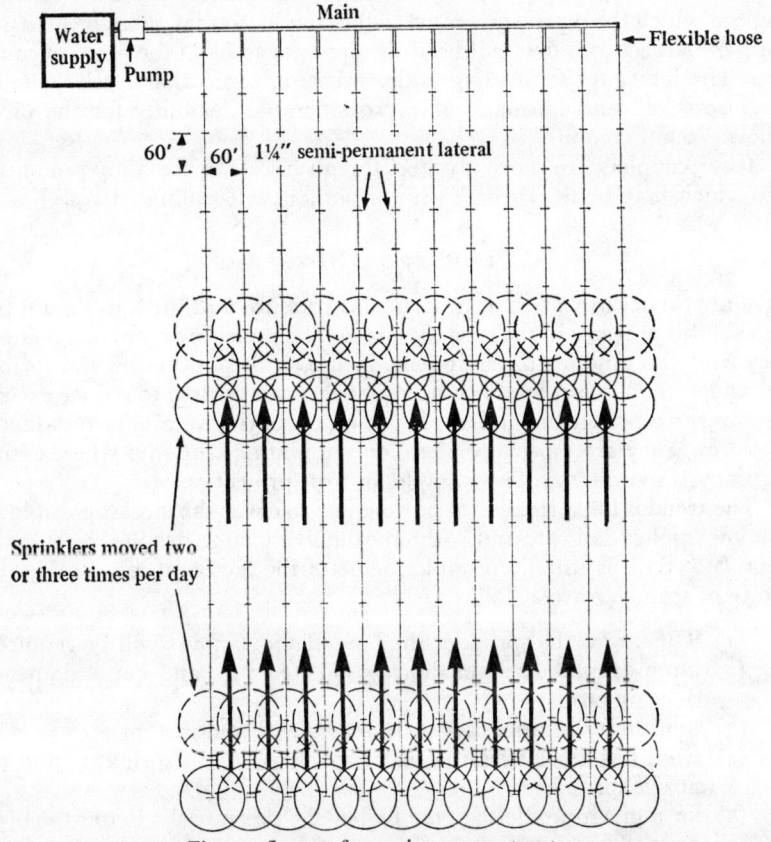

Fig. 12. Layout for semi-permanent system

in the irrigation season. Standpipe connectors are provided with automatic valves as described in the previous paragraph.

Two sprinklers on standpipes are used per lateral, the first being connected in the first standpipe connection and the other halfway down the line. After the desired application each sprinkler and standpipe is moved to the next position. With this system labour requirements and disturbance to the crop are minimal; the latter is particularly important where frequent light applications of water are desirable. These advantages have to be balanced against the heavier capital cost of this system.

SEMI-PERMANENT LAYOUTS WITH AUTOMATIC CONTROL

A further development in the replacement of labour by capital equipment consists of a full complement of nozzles arranged in groups with an automatic valve controlling the water flow to each group. With a sequence controller (see page 63) the valves can be controlled in turn to apply the desired quantity of water to the whole area without a high rate of water supply.

Such systems are suitable for small areas of very intensive horticultural crops.

COUPLINGS

Most portable irrigation pipes are fitted with quick connectors which allow operators to connect pipes easily. Considerable tolerance is allowed for the angle at which the pipe is presented to the coupler so that it can be held at the point of balance and does not need the operator to be at the point of connection. The joints are self-sealing under pressure, come apart easily when the pressure is off, and normally allow considerable flexibility for the pipe to follow ground contours.

Riser couplers are normally steel thread inserts in the alloy pipe fittings into which may be fitted risers with or without self-sealing valves.

FLOW FROM NOZZLES

Each nozzle performs best at a particular pressure and the aim should be to provide this pressure in the sprinkler lines. The pressure at any one point will vary with the amount of equipment in use, with position on the sprinkler line and with topography. Pressure equalizers can be used to reduce excessive pressures to a predetermined level but not, of course, to raise below-standard pressures. They are especially useful on undulating land and where extreme accuracy is required as, for example, in frost protection.

The trend is for sprinklers to be designed to cover the largest possible area at a low application rate and without excessively large droplet sizes; spacing at 60 ft × 60 ft is usually possible. Some of the factors to be considered in choice of sprinkler are as follows:

(1) The larger the sprinkler the fewer lateral moves will be required;
(2) large droplet size can destroy soil structure and cause damage to delicate crops;
(3) uniformity of application decreases with increase in area covered;
(4) wind has more effect on wider spacings, larger sprinklers and twin nozzle sprinklers;
(5) for a given nozzle size, the higher the pressure the better the break-up and the more uniform the application pattern within certain

EQUIPMENT (OUTDOOR)

limits. Excessive pressure can reduce jet throw by causing misting; and

(6) precipitation rates must be kept within the absorption rate of the soil.

Operating pressures for different nozzle sizes

Nozzle size range		Adequate break up pressure p.s.i.	Preferable working pressure p.s.i.
in.	mm		
$\frac{1}{8}-\frac{3}{16}$	3·0– 4·5	30	40 to 50
$\frac{3}{16}-\frac{1}{4}$	4·5– 6·0	40	50 to 60
$\frac{1}{4}-\frac{3}{4}$	6·0–19·0	50	60 to 70

Precipitation Patterns

The distribution pattern from one sprinkler is triangular in shape, with the heaviest application at the centre and a decline towards the outside of the circular wetted area. The geometric pattern of a sprinkler is affected by operating pressure (Fig. 13).

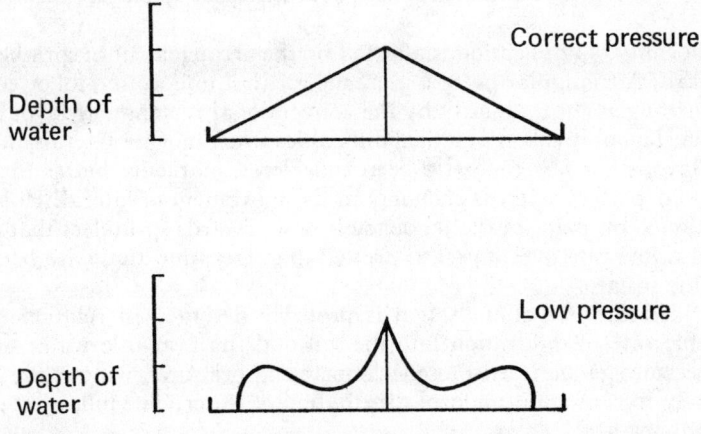

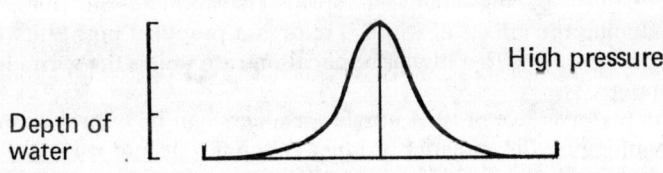

Fig. 13. Effect of pressure on the geometric pattern of a sprinkler

It follows that a considerable overlap is required to get a uniform application of water over an area. This overlap needs to be greater for windy areas, and general recommendations are set out below. Spacings are generally stated as a percentage of the wetted diameter.

Wind space	Lateral spacing
No wind	65 per cent of wetted diameter
5 mph or less	60 ,, ,, ,, ,, ,,
5–10 mph	50 ,, ,, ,, ,, ,,
Over 10 mph	30 ,, ,, ,, ,, ,,

Fig. 14 indicates the uniformity of application that can be obtained with correct overlapping.

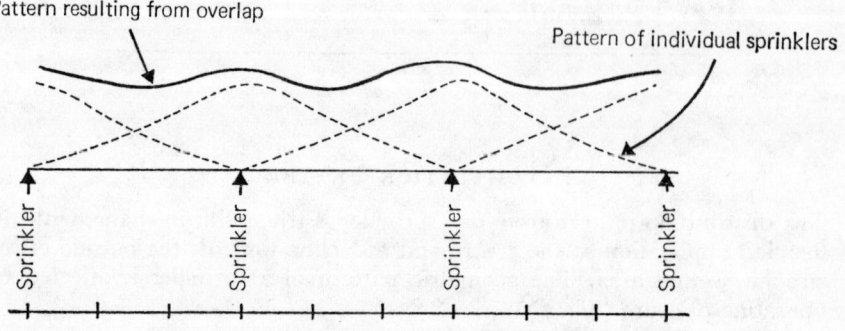

Fig. 14. Distribution of water from overlapping sprinklers

Uniformity of application is affected by the arrangement of sprinklers on the laterals. A triangular pattern gives the greatest uniformity, followed by a square arrangement and lastly by the conventional rectangular layout. The triangular layout involves practical difficulties when moving laterals, and the square layout has not generally been considered markedly better than the rectangular pattern with its economy in the movement of lateral. However, as mentioned on page 43 the tendency is now towards sprinklers that apply water at a low rate over a greater wetted diameter, and these are being set out on the square.

Provided the irrigation system is properly designed in relation to the acceptable rate of infiltration into the soil and the available water supply, and a pressure gauge is used to check operating pressures, wind is the factor most likely to cause non-uniform distribution of water. The following points are of importance:

(1) Wind speeds above 8 mph have a decided influence on water distribution;
(2) reducing spacings between sprinklers and lateral lines, while reducing the effects of wind, is seldom a practical possibility on the farm as it leads to a higher application rate unless the sprinklers are changed;
(3) the performance of twin nozzle sprinklers can be improved in wind by plugging the spreader or inner sweep nozzle and using the range nozzle only;
(4) the height of the riser is important, and a reduction to 18 or 24 in. in height will improve performance in wind.

PUMPS AND PUMPING

Types of Pump

Irrigation water may have to be pumped from a surface supply, i.e., from a river or reservoir, or from below ground, i.e., from a well or bore-hole, and the type of pump should be chosen carefully for the particular installation. A pump may be fixed or mobile and it may be driven by an electric motor, through a tractor p.t.o. or by its own internal combustion engine.

CENTRIFUGAL PUMPS

Centrifugal pumps are basically simple in design and therefore relatively inexpensive to buy and maintain. For these and other reasons the centrifugal pump is generally the most suitable type for irrigation schemes although other types of pump are sometimes used for special circumstances. Centrifugal pumps consist of a scroll shaped casing enclosing a multi-bladed impeller which during operation rotates at high speed. The inlet-pipe connection is on the axis of the impeller and the outlet is at right angles to it (Fig. 15). At a high speed of rotation water is thrown off the tips of the impeller blades, collected by the casing and directed to the outlet. As this occurs water from the inlet-pipe enters the centre of the impeller.

In order to work and to provide lubrication centrifugal pumps must be full of water before starting, and should never be started or run dry. Where the pump is to be above the level of the water supply a hand-operated priming pump (Fig. 9) is often fitted to exhaust air and thus allow the suction hose and pump to fill with water. A foot valve at the bottom of the suction pipe prevents water draining out from the pump and suction line when the pump is stopped and thus eliminates the necessity for re-priming every time the pump is stopped. Specially designed self-priming pumps are available but except on small schemes these are not widely used because of the cost and the lower overall efficiency of the pump.

At constant impeller speed the discharge from a centrifugal pump decreases with increasing delivery pressure, the actual discharge/pressure relationship varies with different models and types of pump. Thus within limits the centrifugal pump can accommodate itself to changes of pressure without excessive loss of efficiency. Nevertheless to obtain the most efficient pump it is essential to specify the maximum and normal pumping duty. The pumping duty is expressed in terms of flow rate at a specified pressure head. The head is the sum of the suction lift, the delivery head (i.e. the height of the highest discharge point above the pump), the friction loss in the pipework at the required flow rate and the desired residual pressure (i.e., the pressure required at the nozzles).

The self-regulating characteristic also allows valves to be shut off for short periods without adverse effects while the pump is running, e.g., when moving lines. The discharge pressure with all the valves closed will be at a maximum and the whole system between the pump and the valves must be capable of withstanding this pressure. Alternatively a pressure relief valve can be fitted in the delivery pipe from the pump. The power requirement of some centrifugal pumps is greatest at maximum flow and no head. Hence to avoid damage to the pump motor by a burst pipe it is safer to install a pump with non-overloading characteristics.

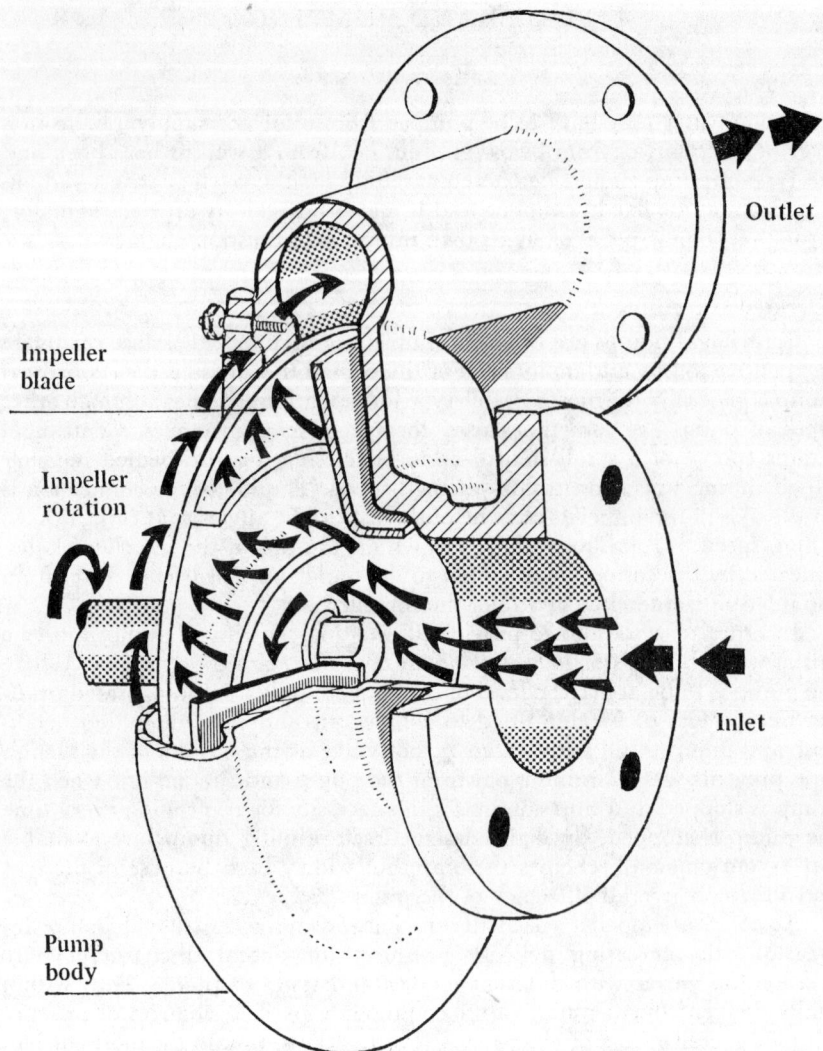

Fig. 15. Centrifugal pump

Single stage centrifugal pumps are available for many farm irrigation schemes, i.e., where the total head is below about 200 ft of water, the actual head being determined by the size and the detailed design of the pump. For higher pressures than can be obtained efficiently from a single-stage unit; multi-stage pumps are required. These have a number of impellers on the same shaft with each impeller working in its own casing. Water from one stage is directed to the inlet of the next, pressure being increased at each stage. In this type of pump clearances between moving and stationary parts must be very fine and small amounts of trash such as straw or stones will cause blockages or damage so that whenever possible it is sensible to choose the cheaper and simpler single-stage version. The maximum suction lift which can be obtained with a centrifugal pump is specified by the manufacturer, but where this is not known it should be limited to not more than 17 ft including suction friction losses. Where the water level is at a greater depth below the ground than this, one of the special pumps now described will be required.

SUBMERSIBLE PUMP

This consists of a multi-stage centrifugal pump directly coupled to an electric motor with the shafts in line to form a unit of sufficiently small diameter to be lowered down a borehole. The electric motor is designed for submerged operation. Submersible pumps are easy to instal and withdraw for maintenance. This type of pump is capable of delivering water at high pressures from considerable depths.

AIR-LIFT PUMP

This pump, which has no moving parts in contact with the water, is particularly suitable where the supply contains sand or grit. It consists of an air compressor sited at ground level and a pipe which delivers air under pressure into an open-end rising main which terminates near the bottom of the well. The aerated water being less dense than the water in the well is forced upwards and collected in a reservoir where the sand can settle out before the water is pumped into the irrigation system by a separate pump.

EJECTOR OR JET PUMP

This is another type of deep well pump with all the working parts at ground level. As is the case with the air-lift pump very high efficiency cannot be expected from ejector pumps. Some of the water from the pump is taken under pressure down the well and discharged in a high speed jet through a venturi at or near the bottom of the rising main which extends below the well-water level. Well water is entrained with the jet and forced up the rising main to the pump.

PISTON PUMPS

Piston pumps are more efficient than centrifugal pumps but are less suitable for irrigation work because at any one speed they deliver a constant quantity of water. This can produce excess pressures and possible damage to pipelines and fitting if there is an obstruction to flow such as the accidental closure of a valve. A pressure relief valve should always be provided with this type of pump.

Siting of Pumps

A pump should be placed as near as possible to the water supply so that the suction pipe is as short and direct as possible. For centrifugal pumps the suction lift should not exceed about 17 ft; greater lifts result in considerable reduction in output. The suction pipe should always rise towards the pump and it should be free from intermediate high points in which air pockets can remain during priming. After the pump starts such air pockets can cause the pump to lose its prime. The end of the suction pipe should always be fitted with a strainer, the total area of holes in this being three times the cross-sectional area of the suction pipe. Where weeds are present in the water additional screening is necessary; this may take the form of a wickerwork or steel mesh basket surrounding the strainer on the end of the suction pipe. A floating strainer should be used when the beds of rivers or ponds are muddy, but the end of the pipe must be sufficiently below the water level to prevent the pump from drawing in air. It is particularly important to check that there are no air leaks into the suction side of the system as even a small leak can prevent a pump from working.

Power Units

The choice of the most suitable power unit to drive the pump depends upon three principal factors, whether a permanent or mobile unit is required, the cost of providing a suitable electricity supply to the site and the importance of keeping capital outlay to a minimum.

A fixed pumping unit is the obvious choice if the water is to be pumped from one point, as from a reservoir or bore-hole. However, where water is being taken from a water course running through the area to be irrigated, it may be possible to extract at several points with a mobile pump thus reducing the length of mains required and the capital cost of the scheme.

Electric motors or internal combustion engines can be used to drive stationary pumps but electric motors should be chosen wherever practicable because of their low maintenance requirements, the ease of automatic control and the simple drive arrangement which can be adopted with centrifugal pumps. Electrically driven pumps are relatively cheap to buy but this advantage may be lost by having to bring the power supply a long way. Generally the best location of the pump-house in relation to an existing suitable power supply is the deciding factor in the choice of the power unit.

Portable pumps are available as units with their own engines or they can be driven from a tractor via the power take-off. The use of a tractor saves capital but the tractor cannot be relied upon for other work during the summer, and furthermore continuous operation at fixed speed is far from an ideal duty for a tractor engine. Because of this it is advisable not to load the tractor at more than about 60 per cent of its belt horsepower and not to use speeds above the maximum recommended for continuous operation. Diesel pumps with **purpose-built** engines are normally designed for continuous running. All diesel engines used for pumping, including tractors, can and should be protected with safety devices to stop the engine if there is a cooling or lubricating failure.

MANAGEMENT OF THE SYSTEM

However well an irrigation scheme has been designed and constructed, to give satisfactory results it must be operated properly. Equipment must be operated only within the design limits. Systems have a maximum capacity and it is important to make certain that in the initial design the system can meet peak demands and is capable of any foreseeable expansion. Addition of extra equipment at a later date may overload and damage the pump. Irrigation equipment must be carefully maintained; a failure during irrigation can result in a considerable loss of benefits or perhaps complete crop failure.

Layout of the Pipework

Main lines should be laid up and down major slopes while sprinkler and lateral lines should be laid across the slope to avoid excessive pressure variation. When pipes are being laid out or moved it is good practice to open the control valve very slightly to allow water to flow along the pipe in order to flush out any soil or debris. After the last length of pipe has been laid and water allowed to flow for a short time through the open end, the end-plug should be put in place. Where more than one distribution line is used, more economical operation is achieved by spacing them along the main and particularly avoiding bunching the distribution lines together at the end of the main farthest away from the pump (Fig. 12, page 51). Sprinkler heads should revolve slowly, and for uniform application it is important to make sure that the risers feeding the sprinklers are vertical.

Starting the Pump

Before starting the pump it is essential to ensure that both the engine and pump are properly lubricated and that water-lubricated pump glands are free.

The valve in the output side of a centrifugal pump must be closed for starting. The pump and suction line should be completely filled with water and primed before the pump is started; failure to do this can damage the pump. Pumps are usually primed by exhausting the suction side with a special hand-operated pump and the manufacturer's operating instructions should be followed. When the pump is running, the valve on the discharge from the pump should be opened slowly to fill the main line and laterals. Rapid opening should be avoided as this could result in water hammer and possible damage to the pipe and fittings.

Operation

Water must never be applied to the crop more rapidly than the soil can absorb it. The pressure gauges fitted on the delivery main from the pump and at the beginning of sprinkler or spray lines or on rain-guns should be checked frequently during operation to make sure that the system is working properly. A variation of pressure from the normal gives an early indication of faults developing in the system such as partial blockage of filters. Variations in pressure affect the application rate and the uniformity of application. If the operating pressure is known, the application rate for the equipment concerned can be determined directly from manufacturer's data. Thus operating at the correct pressure allows amounts of application to be controlled by time alone.

Evenness of distribution is affected by wind and in extreme cases the variation produced can affect crop yields (see page 73 *et seq.*).

Drainage valves must be provided at all low points of a permanent installation so that the pipes may be emptied, either as a frost precaution or for working on the system. The lid of a hydrant box affords little or no protection against frost and pipes and fittings inside hydrant boxes are liable to frost damage. It is wise to empty the system when there is risk of frost or to lag the pipes and valves with sacking or straw before frost occurs. Lagging is not effective against severe and prolonged frost. When draining the system all the valves should be opened to admit air and to prevent pockets of water remaining; the valves should be left open while the system is empty. When portable pipes are not being used they should be stored carefully, preferably under cover and off the ground and stacked with wooden spacers. Sprinkler heads, nozzles and rubber gaskets are subject to wear and should be replaced as necessary. All parts of equipment should be inspected before being put into storage. Rubber gaskets should be removed and stored in water; gaskets of other materials should be treated in accordance with manufacturer's recommendations. Rubber hose must be kept free from frost in a dry place.

Equipment (Indoor)

METHODS OF DETERMINING REQUIREMENT OR CONTROLLING APPLICATION

WATER loss in a glasshouse depends upon the amount of the sun's energy received in the structure and the amount utilized by the crop for transpiration. In the early stages of a crop there is only partial leaf cover and hence only partial interception of the solar radiation. As a crop develops, leaf cover increases until there is complete cover when in theory there will be no patches of bright sunshine on the soil or floor. It can be said that there is complete cover when there are occasional patches of sunlight, depending on the training method adopted. Tomatoes about 4 ft high normally give complete leaf cover. Leaves of commercial horticultural crops utilize about 70 per cent of the solar radiation which they receive so with complete cover up to three-quarters of the incoming energy is utilized. Continuous measurements of the total solar radiation received are made at some Experimental Horticulture Stations and this information can be obtained weekly. The crop's water requirement for the previous week is given and this is an accurate guide to the water application in the following week.

The equipment used on Experimental Horticulture Stations is not practicable for nursery use, but simpler and cheaper, though reliable, intergrating photometers are now available. They operate independently of a power supply, can be calibrated to give a direct reading of water requirement and could be a sound proposition on some of the large glasshouse nurseries.

An evaporimeter is now available of simple and cheap design, readily assembled on the nursery from standard laboratory components. The amount of water indicated by this instrument to be necessary to bring the soil moisture to near field capacity pertains only to the house in which the instrument is located.

When the amount of radiation and crop cover is known it is possible to calculate the water loss in a given glasshouse over a period. Allowances may have to be made for glasshouses with poor light transmission. Where radiation data are not available the water requirements can be estimated from weather observations and Fig. 16.

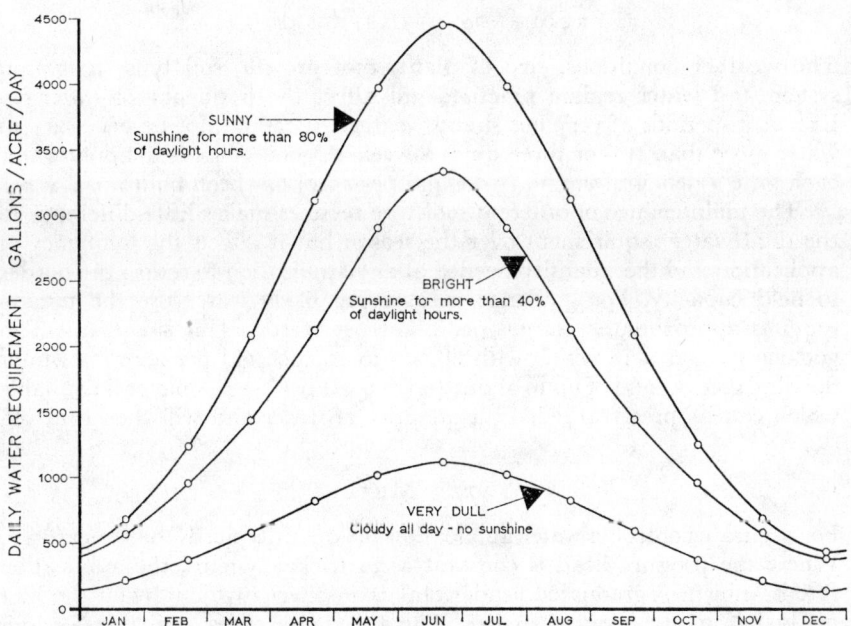

Fig. 16. A guide to the daily water requirements of crops in North–South glasshouses—full crop cover

When this method is used it should be noted that:

the figures given are mean values and should not be used in very poor light areas, e.g., around northern industrial towns; and

when there are long periods of high humidity, e.g., on a foggy day in November and little artificial heat is applied, the daily loss may be reduced to half the values given.

An alternative method of estimating the water requirement is to measure the amount of water in the soil by measuring soil moisture tension with a tensiometer. A tensiometer consists of a porous pot directly connected to a vacuum gauge. With variations in soil moisture tension, water diffuses into or out of the pot resulting in changes of vacuum which are indicated on the dial. A soil at field capacity gives a reading of about 3 cm of mercury and this increases as the soil dries out. The tension at which water is applied depends upon the water regime adopted. With a wet regime, water will be applied at a relatively low tension, e.g., 6 cm, while in a dry regime water will be withheld until a higher tension, e.g., 12 cm is reached. To prevent gross errors two or preferably three instruments should be used, with each porous pot buried between plants with its centre about 6 in. below the soil surface.

Other methods are available but whichever is employed frequent visual checks should be made of the crop and soil in the plant root zone. Plants near the outside of a glasshouse or at the sides of beds require slightly more water than the remainder.

Frequency of Application

The weather conditions, size of plant, root growth, soil type, irrigation system and water regime practised, all affect the frequency of watering. Except in periods of very hot sunny weather, it should not be necessary to water more than two or three times a week. A good guide is to apply water each time a deficit of 0·2 in. (5,000 gal per acre) has been built up.

The maintenance of different moisture regimes makes little difference to the total water requirement over the season but it affects the frequency of application and the quantity needed at any application to return the border to field capacity. For guidance in assessing likely water use the tomato requires approximately 22 gal per plant per season. This is equivalent to 300,000 gal per acre which with 66,000 to 100,000 gal per acre for winter flooding gives a total of up to about 400,000 gal per acre. A meter is available which can be preset to deliver a quantity of water and will then shut off.

Water Meters

For accurate control of water application the quantity needs to be measured. Where the pressure head is constant as in trickle systems, the application rate is known. A graduated header tank is an alternative but by far the best method is to use a water meter. The two main types available for this application are 'rotary piston' and 'inferential'. Both are suitable but the inferential type generally gives less trouble and requires less maintenance. Inferential types must be installed in a horizontal pipeline whereas the rotary piston type can be fitted in any position. Meter size depends on the maximum continuous throughput required. A meter is available which can be preset to deliver a quantity of water and will then shut off.

Fertilizer Application

Fertilizers can conveniently be added to the irrigation water by dilution of concentrated stock solution; this may be done through a number of small dilutors, perhaps one per house, or by a large central dilutor or injection pump. The latter can be a great advantage on a large nursery where a permanent installation under cover with tanks for the preparation and storage of the fertilizers facilitates the control of this important operation.

In all irrigation installations involving the addition of chemicals it is essential to prevent contamination of the public water supply. A storage tank which breaks the direct connection between the mains and the irrigation system fulfills this function. The capacity of such a storage tank should be calculated to maintain a steady supply for the irrigation needs.

Dilutors

Dilutors are of two main types:

VENTURI TYPE

The concentrate is drawn into and mixed with the main flow of water as it passes through a venturi. The equipment is cheap but inaccurate and unreliable and cannot be recommended.

DISPLACEMENT TYPE

The concentrate is displaced from a closed container into the main flow by an equal quantity of water which is bled off and fed into the container. The dilution rate is controlled by adjusting the flow of the displacement water into the container by means of a valve, or by fitting different sizes of jets.

SINGLE AND DOUBLE COMPARTMENT TYPES

In the single compartment type there is no physical barrier between the water and the concentrate, separation being maintained by the differing densities. In the double compartment type the two sections are separated by a flexible diaphram. The single compartment type is the cheaper of the two and provided it is used with reasonable care to avoid shaking and mixing it will give satisfactory results.

Pumps

Both piston and diaphram types of pump may be used for automatic injection of a nutrient solution into irrigation water to give a constant dilution.

Automatic fertilizer injectors provide a dilution rate of from 1 in 100 to 1 in 400; some types may not be variable. Some units incorporate a device which varies the pump stroke according to the water pressure while other manufacturers have adopted a system in which the frequency of operation of a diaphram pump is controlled by the rate of water flow through a meter.

In another system the irrigation water is supplied to the distribution system by a pump drawing via a manifold from a large number of identical pipes of small bore. The desired dilution can be obtained by selecting the number of pipes to be connected to a tank of concentrated nutrient solution in relation to those connected to the water tank.

The electrical conductivity of irrigation water containing nutrients is the basis of another system of maintaining a constant dilution rate. A conductivity cell downstream of the nutrient pump causes nutrient to be injected to maintain a predetermined conductivity in the irrigation water.

Sequence Controllers

These are devices which automatically direct the irrigation water to separate zones according to a predetermined sequence of operations, on a time basis. Manual intervention is necessary only to pre-set the time of irrigation for each zone.

The controller may be arranged to operate entirely automatically, the sensor for water requirement being a tensiostat or an adaption of the electronic leaf used for mist irrigation. More commonly the estimate of water requirement is made by the nurseryman who sets the controller so that the necessary amount is applied.

Sequence controllers are most useful on glasshouse nurseries where the available water supply is sufficient only for a small part of the nursery at one time. A single centrally situated control panel may be equipped to divert water to as many as 100 separate zones in sequence. The extent of any individual zone will depend on the available rate of water supply and the maximum requirement of the crop. Each zone has its water supply controlled by an electric solenoid valve which is energized from the controller. For safety reasons, solenoid valves are operated at low voltage (usually 24 volts). It is convenient to choose valves which permit manual operation (usually in the form of a by-pass) so that the system can be made to work (for inspection purposes) by an operator without going back to the controller to switch on that particular zone.

Sequence controllers may incorporate a number of features. The simplest type has a switch for each zone and the operator selects which zones will be irrigated by setting the appropriate switches. He also sets a single timer which predetermines the time of operation on each zone. In this type of control panel there is thus no method of applying water for differing times at each station. The simplest controllers work on a 24-hr cycle after which they switch themselves off. All controllers have the facility to by-pass automatically any zone where no irrigation is required.

More sophisticated controllers can preselect the length of irrigating time for each station. Different crops at different stages of growth may therefore be watered from one panel. A further stage of development allows the predetermined sequence to be repeated so that all zones may be irrigated twice in each 24 hr and in some units it is possible to programme for up to 14 days ahead. During this period the sequence of operations may be omitted on any predetermined days.

TYPES OF EQUIPMENT

Hosepipe

Watering by hose is still used in some glasshouses but is being rapidly replaced by one or other of the fixed pipe systems. A rose will help to break the impact of the water on the soil but other types of spreader are frequently used. The size of hose should be related to the rate of application and small hoses should not be used where a high rate of application is needed. Hose watering can be used on all crops and for all purposes—ordinary soil watering, overhead damping, and winter flooding. Its main disadvantages are the adverse effect on surface structure and aeration, particularly on soils which cap on the surface, splashing and soil contamination of the lower trusses, the high labour requirement and the difficulty of getting the job done well because of its monotony.

Overhead Spraylines

These can be used in glasshouses to provide the water needs of low growing crops such as lettuce, for overhead damping to relieve water stress, and for winter flooding of border soil. In tomato houses they are commonly used to aid flower setting in the early part of the season for which purpose nozzles forming large droplets are favoured. Overhead spraylines are not suitable for irrigating tall plants with dense foliage, because this causes the water distribution to be uneven. Also, particularly in dull humid weather, the foliage dries slowly thus increasing the susceptibility to disease and making working conditions unpleasant. Flower crops can be watered from above before blooms appear but after this time watering must be at low level.

The equipment should always be made of non-corroding materials. Spraylines are made from aluminium or P.V.C. of $\frac{3}{4}$ in. to $1\frac{1}{4}$ in. internal diameter. Fittings may be of brass, stainless steel, or P.V.C. Joints in aluminium pipe are usually of a proprietary quick release type using a stainless steel clip and butyl rubber seal. P.V.C. pipe may be rigid or flexible. The rigid pipe is usually supplied in lengths having a socket at one end, so the common joint is the spigot and socket, sealed with a solvent cement; butt and sleeve joints are used also, and quick release couplings are available. The flexible pipe is usually joined by means of hose clips and internal sleeves.

There is a wide choice of nozzles and these include three main types: diffuser, slotted and rotary.

DIFFUSER NOZZLES

In this type a jet of water is directed on to a smooth obstruction near the nozzle. Outputs in the range of 4 to 150 g.p.h. can be obtained with a coverage of 3 to 15 ft. The shape of the obstruction and its angle to the jet determine the spray pattern produced. Flat obstructions at right angles to the jet, conical (mushroom head) or hemispherical obstructions with their axes in line with the jets give circular spray patterns, whilst flat or curved obstructions at other angles to the jet produce a fan type of spray. Some of the circular spray nozzles can be adjusted to vary the size of the droplets and type of spray produced.

Pressures used in overhead sprayline systems vary from 10 to 40 lb/sq. in. An increase in pressure at any nozzle reduces the droplet size and increases the discharge rate and the throw, but these changes are not in proportion to the pressure change. A small nozzle with a relatively high water pressure is needed for fine mists. There is always a loss of pressure along a sprayline from the supply end, caused by friction; there may also be a loss due to the slope of the house. The loss which can be tolerated restricts the length of a sprayline.

Some of the mushroom-headed types have a range of heads of slightly different outputs, so that with careful selection the output from all nozzles can be made almost equal in spite of some pressure loss in the sprayline. Circular pattern nozzles are employed to water a relatively wide band from a central sprayline, but fan diffuser nozzles are more suitable for watering bands at the sides of the sprayline.

SLOTTED NOZZLES

In these, jets issue from rectangular or nearly rectangular slots resulting in flat fan jets. The slot may be moulded in a ceramic material or cut in metal. Moulded nozzles are cheap to produce, are available in a wide range of outputs and are capable of being produced to a high degree of uniformity.

The saw cut type of nozzle usually has a high water output and in some, four saw slots are combined in one nozzle. Such nozzles apply water at a very high rate (70–100 g.p.h.) and require large diameter spraylines to avoid excessive pressure drop.

The overall distribution of water depends upon the characteristics of a particular nozzle, the arrangement and spacing of the nozzles and the operating pressure. The spray pattern from different nozzles varies considerably and no general recommendation can be given.

It is often difficult to clean blocked nozzles when the house is fully planted and it is wise to take precautions to prevent blockages. Blockages can be caused by dirt, pipe scale or, in hard water areas, by scale formation due to continued evaporation from the nozzle surface. If the cleanliness of the water supply is in doubt, a filter at the inlet to the line or individual nozzle filters should be employed.

ROTARY NOZZLES

Where a wide coverage of 30 ft diameter or so is required the continuously rotating water driven sprinkler can be used. This applies 60 to 150 gal per hour. These are solely for aiding tomato flower setting, damping down and increasing humidity in glasshouses. They do not give a sufficiently accurate distribution of water for irrigation and should not be used for watering crops.

Low-Level Spraylines

Low-level spraylines are used for applying water to the soil without wetting too much foliage. They are commonly used for tomatoes, cucumbers, carnations and roses, on border soils, straw bales or bed-planting systems. The equipment is sometimes identical with that used for overhead spraylines.

Mushroom head diffuser nozzles and fan jets are common, at spacings of 2 ft to 5 ft. Hairpin type diffuser nozzles are rather bulky and therefore liable to damage when used at low level in a crop. Fan jets providing a semicircular spray pattern are obtainable in brass or plastic. If a smaller angle of distribution is required, moulded plastic nozzles must be used; these can be obtained to produce a spray pattern of about 120 degrees and they usually throw 2 ft to 3 ft. Fan jets can be arranged to provide a lower discharge rate and a smaller spread than anvil jets and, especially if narrow angle types are used, most of the water can be confined to the bed or border soil, thus keeping the paths dry.

Low-level spraylines may be simply supported a few inches above the soil on wooden pegs or suspended from the glasshouse structure. Rigid and flexible P.V.C. and aluminium pipe are all suitable. The flexible types of plastic tube require frequent support to prevent sagging and must be constrained from twisting. Taut wires suspended at about 10 ft intervals to support the spraylines and maintain the desired angle of the nozzles make a good system. Automatic tensioning of the supporting wires and the spraylines by springs or weights is useful to counteract thermal expansion which in P.V.C. is twice that of aluminium.

Adjustment of the throw of nozzles giving semi-circular or narrow angle spray patterns can be achieved by rotation of the sprayline. For watering cucumbers on straw bales, a vertical discharge can be obtained from a nozzle projecting horizontally from the sprayline.

SYSTEMS OF IRRIGATION

Trickle Irrigation

In this method water is emitted at a slow rate from a large number of nozzles which either lie on the soil surface or are supported close to it. There may be one nozzle for each pot plant, tomato plant or cucumber plant, or one for

EQUIPMENT (INDOOR)

up to 12 carnation plants. The basic system consists of a harness of small diameter flexible tubes fitted with nozzles at a predetermined spacing and laid out so that there is a nozzle near each plant (Fig. 17). Various types of nozzle are employed to provide the necessary resistance to limit the flow generally to less than $\frac{1}{3}$ gal per hr per nozzle. A common type consists of a cap which fits loosely over a screw thread to form a compact fine spiral tube of the required resistance. A more recent introduction is the thin-walled plastic tube of 1 in. to $1\frac{1}{2}$ in. diameter which is supplied stitched in the flat state with rot-proof thread close to one edge. Water is supplied at low pressure to the stitched tube through a suitable length of small bore restricting tube from a header, and emerges through the stitching. At the low rate of flow obtained, water drips on to the soil and spreads laterally and downwards below the surface to give a cone of moist soil. The amount of spread varies with the soil type and the rate and frequency of application.

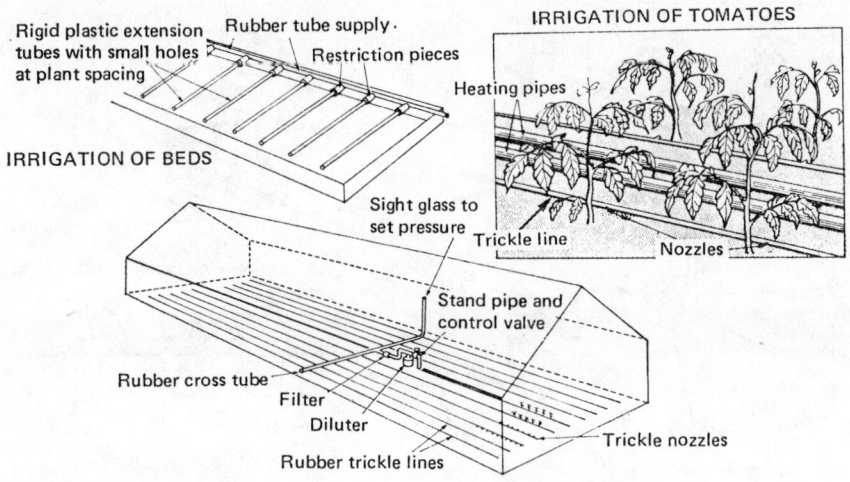

Fig. 17. Trickle irrigation

A good lateral spread can be obtained on medium and heavy loams with the nozzles spaced about 15 in. apart, when the soil a few inches below the surface and between the nozzle can be expected to be uniformly moist. On light sandy soils lateral spread is poorer and unless a close nozzle spacing is adopted zones of dry soil develop between nozzles (Plate X).

A common system for carnations consists of pipes of flexible plastic at frequent intervals across the bed. Several nozzles are buttoned into these transverse pipes which are supplied from a longitudinal pipe at the side or middle.

Water is supplied at a pressure of a few feet head of water from a main which in turn is fed from a header tank, or where there is little fluctuation of the mains pressure, directly from the mains. The pressure head is adjusted by a hand valve. With the very fine passages in trickle nozzles blockages can easily occur and a suitable filter must always be used. Also, the nozzles must be cleaned when necessary to remove deposits of salts arising from liquid feeds or from natural hardness.

Drip Watering

This system is similar to trickle irrigation, but the resistance required to give a low-rate of water application is provided by lengths of small-bore plastic tubes.

In one common system, the 'nozzle' consists of a piece of thick-walled plastic tube of about 1 mm bore and length 1 to 4 ft; one end is inserted into a hole in the distribution tube and the other is fixed in a plastic peg for firm support (Fig. 17 and 18). Another similar system uses 3-mm bore tube, which is further restricted by a plastic grooved rod, which also locates and supports the tube. An advantage of this system is that the resistance can be adjusted by using different lengths of resistance tube or rod to compensate for other factors, such as slope and pipe friction.

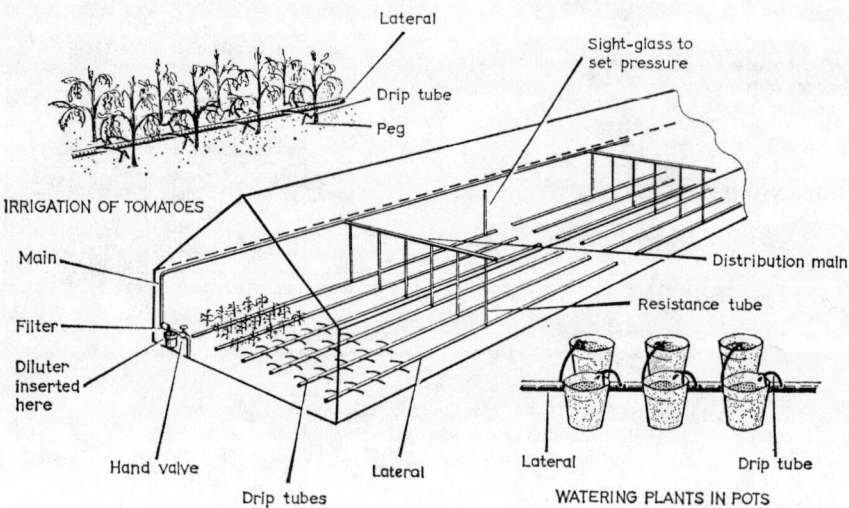

Fig. 18. The drip watering system

In some applications the use of flexible final resistance tubes can make it easy to move plants in pots and to modify patterns of water distribution on beds and borders, but sometimes the tubes can get in the way. When hard water is used deposits which are difficult to remove can form in the resistance tubes; this can be prevented by adding small amounts of dilute nitric acid to the water, but advice should be obtained on the amount to add. The initial cost of this system is slightly more than that of a trickle system.

Low-Level Sprinkle System

This system uses small-bore rigid P.V.C. ground-level pipes drilled with small holes from which the water issues as a fine jet. The trajectory of the jet can be varied by adjusting the pressure head with a hand valve. The system was designed to give a larger volume of wetter soil than a trickle irrigation system. It gives a very even distribution of water on level sites but unless precautions are taken this rapidly becomes less uniform as the slope increases.

EQUIPMENT (INDOOR)

LAY-FLAT TUBE SYSTEM

This system employs thin walled lay-flat polythene tubes punched with small holes. The tubes are laid down the centre of beds or between double rows of plants, one end being stopped and the other connected to a distribution main (Fig. 19). When the water is turned on the tube fills and becomes roughly cylindrical, with jets of water issuing from the holes. As with the low-level sprinkler system the throw of the jets can be altered by using the stand-pipe tap to adjust the head of water. The method is considerably cheaper than other systems but may require more attention. It puts on water at a very high rate and the application is not very uniform. With four $\frac{1}{32}$ in. holes every 5 in. the rate of application is so high that it restricts the length of tube that can be operated at one time and puts a very heavy demand on the water supply. Careful planning can minimize the effect of slope, but because of the low-operating head and the large amount of water contained in the tube, which drains to the lower end each time the system is turned off, water is less accurately distributed than with other methods. The method has been mainly used on tomatoes and carnations and is extremely effective for winter flooding.

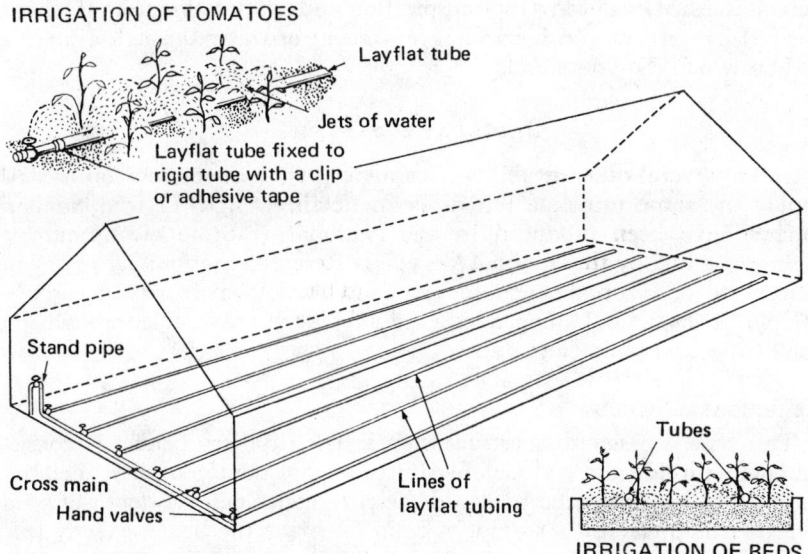

Fig. 19. The lay-flat tube system

CAPILLARY WATERING

This term is used to describe the automatic supply of water to plants growing in containers on moist sand by capillary movement through the compost from a constant level water-table in the sand. The water-table is controlled at a level below the container base so that the compost is never waterlogged. This system can be employed on raised benches or in beds at ground level. It eliminates most of the labour of hand watering and supplies each container according to its individual water requirement; thus plants of different species and at different stages of growth are adequately watered. Many plants thrive on capillary watering but maximum benefit is obtained with fast growing plants having a high water requirement. This system is not suitable for all plants, particularly those requiring a dry regime, and before large scale production is undertaken a trial should be carried out with the particular species.

The principle of the system is that water will move upwards above the free level into small soil pores by capillary attraction, the rise being greatest for the smallest pore sizes. In composts there is a range of pore sizes and the capillary rise therefore varies; by selecting a suitable compost and free water level in relation to the pot, the smaller pores will be filled with water while air will remain in the larger ones. Hence the water/air ratio and the moisture content of a compost varies at different levels in a pot, the compost at the top being rather less moist than that at the bottom. If the free water level is too near the plant roots or the compost contains all small particles, all the spaces will be filled with water and an anaerobic condition will result. On the other hand, if the free water level is too far below the plant roots or the compost is composed of large particles, insufficient water will be supplied to the plant. Thus the vertical distance between the free water level and the pot is critical for successful watering by this method. Once capillarity has been established, water loss by transpiration and evaporation from the pot is replaced by vertical and horizontal movement of water through a layer of sand on which the pots stand.

Capillary Systems

There are several different forms of capillary bench in commercial use. All employ the same principle but differ in detail. Designs of true capillary benches have been produced by the National Institute of Agricultural Engineering and by the National Vegetable Research Station. These designs incorporate a horizontal bench, the sides and base of which are waterproofed with plastic sheet, sand filling, a longitudinal pipe or gravel-filled trough, and a ball valve and tank. (Fig. 20).

THE IRRIGATED BENCH

This type is easier to construct than a true capillary bench. It consists of a base with side and end members fitted so that the top of these members is 1 in. to 2 in. above the base; this height determines the depth of sand. Though it is preferable, it is not essential to make the bench exactly level provided that sections are kept short, e.g., by pieces of upturned plastic film (Fig. 21); watering is less uniform on benches which are not level. No attempt should be made to make the bench hold water, indeed, adequate drainage should be provided at the sides and ends of the bench. Any type of base can be used, e.g., flat or corrugated sheets with the corrugations running in either direction and strong existing benches can be used with little or no modification. This and the lower cost of construction are the principal advantages of this bench compared with the capillary bench. Sand is placed on the bench and struck off with a straight edge across the side members. Water is applied by one or two low-level irrigation lines, for instance by trickle irrigation, laid on the surface of the sand and running the length of the bench. The irrigation system is connected to the water supply and liquid feed equipment.

EQUIPMENT (INDOOR)

CAPILLARY SAND BENCH FOR AUTOMATICALLY WATERING PLANTS IN POTS AND BOXES

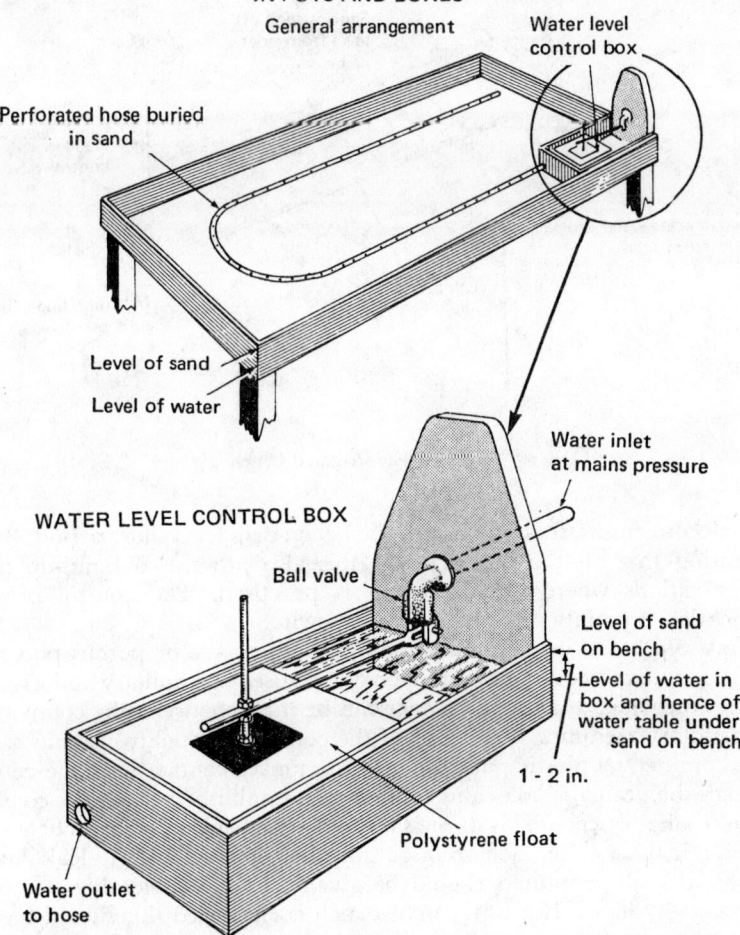

Fig. 20. Capillary sand bench for automatically watering plants in pots and boxes—general arrangement

Several alternatives to the sand surface for the pots to stand on are now available. They take the form of an absorbent mat fixed on one or both sides of a base member which provides resistance to tearing. Use of these materials offers the facilities of a good capillary medium which is lightweight, quickly replaceable, and reduces the amount of sand on the bench. Sand may be needed only for levelling corrugations in the asbestos base. For new bench construction proprietary slotted angle and flat asbestos sheets may be appropriate.

The required frequency of water application depends upon the water holding capacity of the sand or matting and type of sand used and the rate of water loss; in bright weather hourly applications may be desirable. If water is applied too infrequently pots are likely to dry out and plants may root through excessively. When a pot has dried out it may not take up water again when the bench is rewetted. Sufficient water should always be applied at each

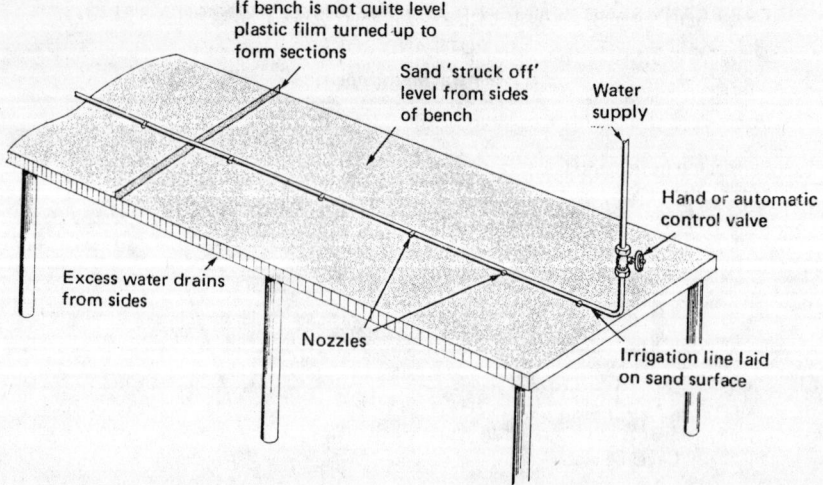

Fig. 21. The irrigated bench

irrigation to ensure that all parts of the bench drip for a short period. Besides indicating that all the bench is wet this also prevents a build-up of salt concentrations where liquid feeding is practised. The control of water application is mainly manual or by time switch.

Any type of plant container such as clay, plastic or porous pots up to about 6 in. high, seed-boxes and so on can be used on capillary and irrigation benches provided that good contact can be made between the compost and the capillary medium. If the base of the container is relatively thin as with most commercial plastic pots, the medium makes contact with the compost through the drainage holes and capillarity is established when the container is placed on the bench. It is unecessary to place a wick in the basal hole unless the base is thick or the hole so large that the compost tends to leak out.

Initially all containers should be given one and preferably two heavy waterings by hand. If a water-tight bench is employed this should be done before the containers are placed on the bench.

After this stage watering will continue automatically but a regular check should be kept for signs of water stress. This indicates that capillarity has broken down and must be re-established.

Mist Propagation

This system of overhead watering supplies moisture directly to the leaf surfaces of cuttings set in compost on a bench. It prevents wilting of the leaves by reducing transpiration.

A moisture detector commonly known as an 'electronic leaf' is often used to control the misting system. This consists of a small plate of insulating material carrying two electrodes; it is positioned amid the cuttings so that it is wetted, like the plant leaves, by water emitted from diffuser misting nozzles. When the electrodes are bridged with a water film, thus completing an electric circuit, the control unit operates a solenoid valve shutting off the water supply. Bursts of water supply may be of a few seconds duration only, and may be made at similarly short intervals. A control unit to vary the frequency of mist applications is available and can be helpful for weaning plants to natural growing conditions.

Special mist propagation nozzles are available, for operating at 25 to 50 lb/sq. in., but simple diffuser nozzles can be used at high pressures to produce sufficiently fine droplets.

The water supply pipe is usually installed in the bench below the rooting medium with suitably spaced vertical pipes, each fitted with a single nozzle. It is very important to ensure even mist coverage of the cutting benches. Overhead misting during propagation is most beneficial when it is used in conjunction with bench warming.

Filters

The quality of the water supply for glasshouse irrigation is important; adequate filters should be installed on the downstream sides of any fertilizer diluting equipment to remove all but the finest solid particles. An 80-mesh screen is suitable and may be made of bronze, stainless steel or nylon. Blockages may occur due to chemical changes in hard water or during liquid fertilizer application. Phosphate may be precipitated from nutrient solutions by salts dissolved in the water; iron salts may be deposited in small-bore tubes.

A few types of nozzle may be easily cleaned by hand, notably the pin type of anvil jet. Many types however cannot be dismantled easily, but may be kept fairly clean by passing dilute nitric acid through the system occasionally. Commercial 70 per cent nitric acid is diluted to 1 part of acid in 2,000 parts of water, and 1 gal of commercial acid is applied to an acre of equipment. The treatment is ineffective against iron salts. *Care is needed when handling concentrated nitric acid.*

Irrigation in Practice

IRRIGATION is used in agriculture for four purposes:

> to provide water for the growth of crop plants by supplementing rainfall; in protected cropping, irrigation replaces rainfall;
> to facilitate cultural practices, such as seedbed preparation or harvesting, by changing the condition of the soil;
> to protect plants from frost; and
> to distribute water carrying chemicals used as fertilizers or pesticides.

The first of these is described in pages 1 to 4. The second uses irrigation water to soften the soil sufficiently deeply to allow cultivation or harvesting to be satisfactorily carried out. Success depends upon applying no more water than is necessary for this and on doing the job as soon as the soil is dry enough to carry the machines but before it dries out again to an unfavourable condition. The last two purposes are described in pages 75 to 76 and on page 102.

IRRIGATION FOR CROP GROWTH

The factors affecting irrigation decisions on when to water and how much water to apply are climate, soil type and plant response. The main climatic considerations have been outlined under Meteorological Aspects; but one practical problem now needs consideration. Because rain may fall immediately after irrigation, resulting in a waste of water and leaching of nutrients, it is common practice on the more retentive soils to satisfy only part of the water requirement, leaving a small soil moisture deficit after irrigation. Such planned deficits are given for specific crops and soils in the crop sections which follow.

Soil type and depth affect its ability to hold water to be extracted and used by the growing plants. For use in irrigation practice, the assessment of available moisture in the soil should be made within the rooting depth of the crop to be irrigated. The limit to rooting depth may be the character of the particular crop or the inability of the roots to penetrate the soil because of adverse conditions such as impeded drainage, soil pans, rock layers or severe acidity; those can occur in any depth in the soil. If there is any restriction in rooting it will result in the need for more frequent and smaller applications of water than are required by the same crop on a similar soil where depth of rooting is not impeded.

Organic matter also affects the available water capacity of soils, directly as a result of a higher organic matter content or indirectly through an improvement in soil structure.

Very stony soils are less retentive than stoneless soils of the same texture. For this reason all gravel soils should be placed in Group A, Low, Table 1 which gives the soil moisture properties of typical soils.

Plant response is the basis for profitable watering particularly with annual crops whose produce is the fruit. Where crops are grown from seed, water is necessary for germination. Adequate soil moisture at this time assists even germination and also provides a soil reservoir of water throughout the early life of the plant. If the seedbed is dry, irrigation before sowing a crop is generally better than watering a newly sown crop, and the use of small amounts of water in this way can be very rewarding. There may be difficulties on sandy soils as the impact of water droplets on many of these will cause slaking; this may cause capping on drying, preventing emergence and resulting in the killing of seedlings. Also on many soils excess irrigation may produce a temporary exclusion of air from the soil around the seed, causing death by suffocation.

Transplanted crops often need irrigation to aid establishment by providing water for the reduced root system until the growth of new roots enables the plant to exploit a larger volume of soil.

During growth, leaf crops such as grass or brassicas are most responsive to a continuous water supply throughout their life. Nevertheless there may be periods in their life when response to water is greater than at other times. For example a given quantity of water will result in a greater yield increase in cabbage and cauliflower if applied 15 to 30 days before the expected commencement of harvest than if applied when the plants are smaller. With other examples, such as grass, tree fruits or celery, there appears to be no particular time during the summer when watering is more effective than at other times. Seed-producing crops, such as cereals or pulses give the greatest yield responses to water applied just before or at the flowering stage. Earlier provision of water may increase straw or haulm, but not the crop. Where crops have marked responses to water applied at a particular growth stage, this is indicated under the sections describing specific crops.

As a general rule maturity is delayed by irrigation but this is seldom significant except for crops where earliness commands a high premium and even a few days' delay in maturity may result in a severe drop in price. Irrigation may also change quality for better or worse according to crop and circumstances.

There are many economic considerations in irrigation, such as the limitations of soil and climate, the value of the crop, the cost of water, limitation of water supply and the level of management. A specific assessment of likely profit should always be made.

At the present time the experimental evidence on which to lay down precise recommendations is limited and future advice may have to be modified as more information becomes available. The individual crop recommendations are the optima so far as is known and may require more water than is available. Water applications less than these may give lower yields but are frequently justified economically. Application of water in excess of the recommended quantities will not improve returns and will frequently cause yield or quality depressions and also leach plant nutrients from the soil. Excessive application of water should, therefore, be avoided.

With limited water supplies it is usually better to concentrate the whole of the available water on a smaller area of crop and to irrigate this effectively rather than apply a little water over an extensive area (tree fruit crops are an exception to this rule).

Any possible increases in yield mentioned under the various crops are merely examples, as the increases to be expected depend on many factors including the weather, soil type, fertility, the location of the farm, the stage of growth of the crop, and the general standard of farming.

Where crops of low cash value are being grown the benefits of irrigation are likely to be small and probably uneconomic unless water can be applied at a moisture sensitive growth stage, such as flowering in seed-producing crops, when the yield response may be well worthwhile. With high value crops the cost of irrigation can be covered by small increases in yield and these are most likely to give a good cash return.

Terms Used in Defining Irrigation Periods

PRE-SOWING

Before drilling, to ensure that the seed is placed in soil with sufficient moisture for quick germination and to keep the seedling growing quickly until it is large enough for overhead watering without damage. The usual damage is from large droplets falling on to the soil and displacing the air in the spaces between the soil particles. This temporary depletion of oxygen from the soil around the roots kills seedling plants. When the seedlings cover more ground, the leaf canopy breaks the force of the droplets and no damage is done. Pre-sowing watering may be done before basic cultivations such as ploughing; after ploughing, but before working down the seedbed; or after seedbed preparation but before drilling. Exceptionally, as on coarse seedbeds, watering may be done after drilling. After irrigation the length of time required before cultivation or drilling can be done satisfactorily is very variable and must be judged by experience.

COTYLEDON STAGE

When the plant has the first seed leaf or first pair of seed leaves but before true leaves grow.

TWO (THREE) (FOUR) LEAF STAGE

When the plant has two (three) (four) true leaves not counting the seed leaf or leaves.

LATE TILLERING STAGE (OF CEREALS)

When the leaf sheaths become strongly erected but before the first node of the stem is visible. (Growth stage 5 on the Feekes scale).

EARLY FLOWERING OR GREEN BUD STAGE

When the first inflorescences or single flowers are easily seen but before the green sepals open to show the colour of the petals.

WHITE (RED OR OTHER COLOUR) BUD STAGE

When the sepals of the first opening flowers part to show the colour of the petals of the opening flower.

BOOT STAGE (OF CEREALS)

When full stem elongation has taken place and the ear can be felt within the leaf sheath just before emergence. (Growth stage 10 on the Feekes scale).

FULL FLOWER

When approximately 50 per cent of the flowers on each plant are fully open.

POD SWELLING (OF LEGUMES)

When the pod has set and the seeds inside it start to enlarge.

TUBER SWELLING OR MARBLE STAGE (OF POTATOES)

When the tubers at the ends of the stems are 1 to 2 cm diameter in about late June; it coincides with the appearance of the first white or coloured petals on most flowering varieties.

THROUGHOUT LIFE

This term indicates the period between emergence of the seedling and harvesting the crop, with an overall limitation of the calendar period April to August inclusive unless different dates are specified. Soil moisture deficits should be calculated from 1 April or from drilling, whichever is the earlier, assuming that there is adequate soil moisture at the beginning of April or at drilling.

Spinach seedlings in a box showing 'capping' produced by too much water

Boxes of seedlings showing the different results from watering (*left*) from the bottom and (*right*) from using large droplets

PLATE I

Photo: *National Vegetable Research Station*

Tilth damage caused by the impact of large drops during irrigation

Photo: *National Vegetable Research Station*

Lettuce given the same quantities of water as above, but applied as mist. The plants are larger and were slightly earlier

PLATE II

Photo: Wright Rain Ltd.

Conventional layout of sprinklers along a lateral

Photo: Wright Rain Ltd.

Conventional layout with reduced sprinkler complement. The next position for the sprinkler can be seen at the left-hand edge of the photograph

PLATE III

Photo: Farrow Irrigation Ltd.

A sprinkler in the process of being inserted into a self-sealing quick-coupling on a lateral

Photo: Wright Rain Ltd.

A semi-permanent layout showing a small diameter lateral and self-supporting stand-pipe with a quick-coupling

PLATE IV

Photo: Wright Rain Ltd.

Low angle sprinkler irrigating an apple orchard in Kent. The low projectory of the water stream from the nozzles does not wet the foliage or fruit

Oscillated sprayline watering self-blanching celery

PLATE V

Photo: Wright Rain Ltd.

Liquid manure being distributed at 100 gallons per minute through a manure rain-gun

Photo: Farrow Irrigation Ltd.

A grassland version of a self-travelling sprinkler

PLATE VI

Most quick couplings allow a certain amount of out-of-alignment, which is useful on undulating ground or when the portable main is laid along an uneven headland

Photo: National Institute of Agricultural Engineering

A positive non-leak coupling

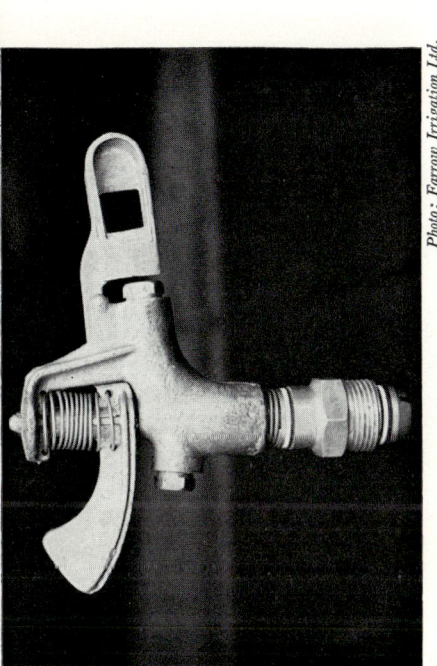

Photo: Farrow Irrigation Ltd.

Single nozzle oscillating sprinkler

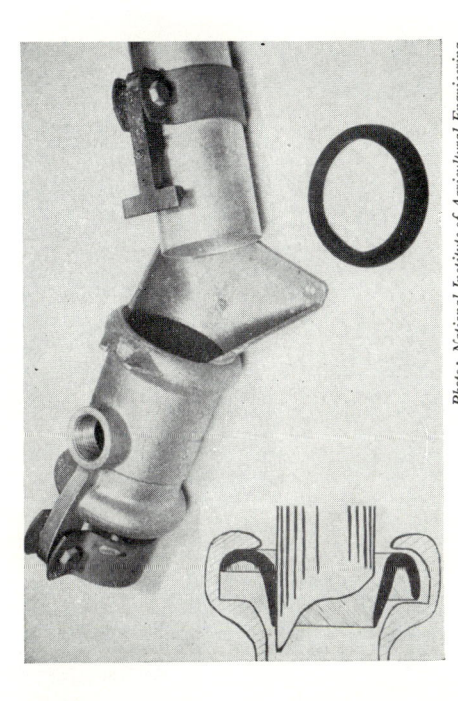

Photo: National Institute of Agricultural Engineering

A self-draining coupling. The diagram shows the rubber washer released. When water pressure expands the washer, it grips the pipe and the outer tube

PLATE VII

Photo: National Institute of Agricultural Engineering

A single stage transportable pump driven by a petrol engine. The floating suction filter keeps the end of the pipe-off the bottom of the pond and far enough below the water surface to prevent the entry of air

Photo: National Institute of Agricultural Engineering

A tractor p.t.-o is a suitable source of power for an irrigation pump if the tractor can be spared. The hand-operated priming pump is used to empty the suction pipe and has a special coupling to prevent air leaking into it

PLATE VIII

Trailer adapted for carrying irrigation equipment

Photo: Wright Rain Ltd.

A recently introduced giant rotary irrigator

PLATE IX

Photo: National Vegetable Research Station

Trickle equipment used in the field

Black currants protected from frost with layer of ice

PLATE X

Photo: National Vegetable Research Station

Butyl lined sand bench showing method of construction and control valve water-level

If there is enough labour available, the moving of the equipment can be speeded up by carrying several sections at a time

PLATE XI

Photo: National Vegetable Research Station

Copious watering (*left*) increased the yield of leaf-vegetable such as cabbage

Photo: National Vegetable Research Station

Summer cauliflower plants grown under cold glass during the winter are usually sparingly watered (*left*). If copiously watered (*right*) they eventually yield bigger curds irrespective of any summer irrigation

PLATE XII

IRRIGATION OF CROPS GROWN IN THE OPEN

Information is given in an abbreviated style where possible, particularly for horticultural crops; certain crops such as Fruit, Hops and Grass are covered in a more extended manner.

FARM ROW CROPS AND VEGETABLES

ASPARAGUS

Prospect—unlikely to increase yields. Up to the end of June there is no water loss from the plant, only from the soil. Later, when the bower grows, the scale-like leaves and needle-like branchlets are well suited to resist transpiration. Irrigation not advised.

BARLEY

For information on barley see under Cereals on page 80.

BEANS, BROAD (FOR MARKETING FRESH AND FOR PROCESSING)

Prospect—Economically marginal. Irrigate only under extreme conditions.

Period—At the early flowering stage, which is between mid-May and the beginning of July according to the planned date of harvesting; also at pod swelling.

Treatment—All soils; 1 in. at early flowering stage unless 1 in. or more of rain falls at this time. On Class A* soils a further inch at pod swelling.

BEANS, FRENCH, DWARF OR STRINGLESS (*Phaseolus vulgaris*, EITHER BUSH OR CLIMBING, GROWN FOR MARKETING FRESH OR FOR PROCESSING)

Prospect—Routine treatment.

Period—Pre-sowing; early flowering.

Treatment—All soils; if the soil is so dry that pre-sowing irrigation is needed, give 1 in. in April, or 2 in. in May. At early flowering give 1 in. unless 1 in. or more of rain falls at this time.

BEANS, RUNNER (*Phaseolus Coccineas*, EITHER BUSH, PINCHED OR CLIMBING CROPS, GROWN FOR MARKETING FRESH OR FOR PROCESSING)

Prospect—First priority. Although in some experiments yield responses have been disappointing, irrigation should be provided for this crop whenever possible.

Period—Pre-sowing; from early flowering onwards until the end of August.

Treatment—All soils. If the soil is so dry that pre-sowing irrigation is needed, give 1 in. at that time. From early flowering stage until end of August on Class A soils give 1 in. at 1 in. S.M.D.; on Class B soils give 2 in. at 2 in. S.M.D. and on Class C soils give 2 in. at 3 in. S.M.D.

---

* See Table 1.

Beetroot (Globe or Long Table Beet, Grown for Bunching or Topping)

Prospect—Routine treatment.

Period—Pre-sowing and throughout life; bunched crops pre-harvest.

Treatment—The pre-sowing application is most important, particularly with crops sown in May and June, in order or get an even plant stand. When required in April, give 1 in. In May or June give 2 in. Post-sowing irrigation is less important except on Class A soils. On Class A soils give 1 in. at 1 in. S.M.D. up to the end of August; on Class B and C soils give 1 in. at 2 in. S.M.D. up to the end of August. Early bunched beet will benefit from 1 in. immediately before pulling to aid lifting if the soil is dry.

Brussels Sprouts

Prospect—To establish transplanted crops only.

Treatment—Spot watering at or after planting in dry weather is adequate. If overall watering is done give 1 in.

Cabbage, Spring Hearted

Prospect—Marginally economic in most seasons and areas.

Period—Pre-sowing; pre-harvest.

Treatment—In late July or early August the soil is often wet enough to drill without watering. When it is not, on all soil types give 1 in. water. In a dry April or May, on all soil types, give one application only of 1 in., 20 to 10 days before cutting is expected to start.

Cabbage, Greens (For Cutting From November to April)

Prospect—Marginally economic.

Period—Pre-sowing.

Treatment—In late July or early August the soil is often wet enough to drill without watering. When it is not, on all soil types, give 1 in. water. Occasionally, in a dry spring, it may be beneficial to irrigate with $\frac{1}{2}$ in. water after applying a nitrogenous top-dressing, but the soil is often too cold for this to be beneficial.

Cabbage, Summer and Autumn

Prospect—Routine treatment. Where irrigation is adequately used, closer spacing is possible, giving a big increase in yield per unit area. Where only a little water is available, normal spacing should be given.

Period—Pre-sowing or pre-planting; throughout life *or* pre-harvest.

(1) All crops. When the soil is dry enough, on all soil types, water before sowing or planting; give 1 in. in April or 2 in. in May or June.

(2) Where adequate water is available and plants are closely spaced, during the growing period, on Class A and B soils give 1 in. at 1 in. S.M.D. or, on Class C soils give 1½ in. at 2 in. S.M.D., up to the end of August.

(3) Where water is limited and normal or wide spacing is used, satisfy the S.M.D. up to 2 in. about 20 days before cutting is expected to start. If the S.M.D. is more than 2 in., give 2 in. water.

Cabbage, Winter and Savoy (Including Dutch White)

Prospect—Marginally economic.

Period—Direct-drilled crops, pre-sowing; transplanted crops, to establish transplants only.

Treatment—When the soil is dry enough, on all soil types water before sowing. Give 1 in. in April or 2 in. in May to July. In dry weather spot-water transplanted crops or, if overall watering is done, give 1 in.

Carrots

Prospect—There has been little response to irrigation in experiments with carrots, but, tentatively, on Class A and B soils it is probable that irrigation will be economic in dry years.

Period—Pre-sowing; during growth up to the end of August. Bunched carrots only, pre-harvest.

Treatment—When the soil is dry enough, on all soil types, water before drilling. Give 1 in. in April or 2 in. in May to July. If irrigation is given during growth, do not water between sowing and the four true-leaf stage. Experimental evidence has shown that plant stand can be reduced by watering before the crop is large enough. After this growth stage, on Class A soils give 1 in. at 1 in. S.M.D.; on Class B soils 1½ in. at 2 in. S.M.D. When this crop is irrigated, do not allow large deficits to build up, or watering may then increase splitting. When the soil is dry, carrots for bunching may be irrigated immediately before lifting to make pulling easier.

Cauliflowers, Early Summer

Prospect—On Class A and B soils, first priority. On Class C soils, routine treatment.

Period—Throughout life of crop *or* pre-harvest.

Treatment—On Class A and B soils, give 1 in. at 1 in. S.M.D. On the better soils this treatment will enable plants to be more closely spaced than normal. On Class C soils, or with widely-spaced crops on Class A and B soils, give one application only, about 20 days before cutting is expected to commence. Satisfy the S.M.D. up to 2 in. If the S.M.D. is more than 2 in. give only 2 in. water.

Cauliflowers, Summer and Autumn

Prospect—On Class A and B soils, first priority for crops harvested in July, and August, routine treatment for crops harvested from September onwards. On Class C soils, routine treatment for all crops.

Period—Pre-sowing; throughout life or pre-harvest.

Treatments—For direct drilled crops on all soil types, if the soil is dry enough, water before drilling, 1 in. in April, 2 in. in May or June. For transplanted crops, spot watering or overall application to aid establishment of transplants. If an overall application is given, give 1 in. For crops harvested in July and August on Class A and B soils, give 1 in. at 1 in. S.M.D. throughout the life of the crop. For these crops on Class C soils, give one application only, about 20 days before it is expected that cutting will start. Satisfy the S.M.D. up to 2 in.; if the S.M.D. is more than 2 in., give 2 in. water. For crops harvested in September on all soil types give one application of up to 2 in. (satisfy the S.M.D. if this is less than 2 in.) about 20 days before it is expected that cutting will start. For crops harvested after September, watering after establishment is unlikely to be economic.

Cauliflowers, Winter (Roscoff and Winter Hardy Types)

Prospect—Marginally economic.

Period—Direct drilled crops, pre-sowing; transplanted crops, after planting to aid establishment.

Treatment—On all soil types, if the soil is too dry at sowing time, water before drilling with 2 in. With transplanted crops spot or overall water throughout life until the end of August.

Treatment—Immediately after planting, give 1 in. Thereafter, during June and July, give 1 in. at $1\frac{1}{2}$ in. S.M.D.; during August give 1 in. at 2 in. S.M.D.

Celery, Self Blanching

Prospect—First priority, particularly on mineral soils.

Period—After planting to aid establishment and throughout life until the end of August.

Treatment—On all soil types, give 1 in. at 1 in. S.M.D.

Cereals

Prospect—Experimental work has shown that cereals will respond to irrigation applied in a dry spring. This situation occurs only occasionally and irrigation of cereals is an opportunity to make use of equipment if it is not needed for use on crops which will give a higher return from irrigation.

Period—Late tillering to boot stage (May to June).

Treatment—On Class A soils apply $1\frac{1}{2}$ in. at 2 in. S.M.D.; on Class B and C soils 2 in. at 3 in. S.M.D.

Leeks

Prospect—Routine treatment.

Period—For direct drilled crops sown in April, pre-sowing; for transplanted crops, after planting to aid establishment; all crops, throughout life until the end of August.

Treatment—On all soil types, if the soil is too dry for drilling, water before drilling. Give 1 in. water. On transplanted crops, water along rows, or give overall application, to aid establishment. If an overall application is made, give 1 in. During the growth of the crop, until the end of August, on Class A soils, give 1 in. at 1 in. S.M.D., on Class B soils 1 in. at 2 in. S.M.D. and on Class C soils, 2 in. at 3 in. S.M.D.

Lettuce, Summer

Prospect—First priority.

Period—Pre-sowing; throughout life until the end of August.

Treatment—On all soil types if the soil is dry at sowing, water before drilling. Satisfy the S.M.D. up to 1 in. in April, or up to 2 in. in May to July. To avoid capping do not water after sowing until the plants have four true leaves. Plant stand has been reduced by overhead irrigation of very small seedlings. After the four true-leaf stage up to the end of August, on Class A and B soils, give 1 in. at 1 in. S.M.D. and on Class C soils 1 in. at 2 in. S.M.D. Where water is limited and it has not been possible to irrigate in this way, give one application of 1 in., 14 to 21 days before cutting is expected to start.

Lettuce, Winter

Prospect—Marginally economic.

Period—14 to 21 days before cutting is expected to start.

Treatment—Give an application of only 1 in.

Marrows

Prospect—Routine treatment.

Period—Pre-sowing (direct drilled crops), after planting out, and throughout life until the end of August.

Treatment—Direct drilled crops, if the soil is dry enough before sowing, give 1 in. in April, 2 in. in May. Transplanted crops, give 1 in. to aid establishment. Thereafter, on all soil types, give 1 in. at 1 in. S.M.D.

Oats

For information on oats see under Cereals on page 80.

Onions, Bulb

Prospect—Marginally economic on most soils. Routine treatment on Class A soils. The uneven application of irrigation can cause uneven growth and ripening, which is undesirable so it is better to grow this crop on soils where irrigation is unlikely to be needed.

Treatment—On Class A soils, give 1 in. at 1 in. S.M.D. throughout life until the end of July. Do not water after this date or the ripening of bulbs may be delayed.

Onions, Salad

Prospect—Routine treatment on summer crops; marginally economic on crops sown after mid-July.

Period—Pre-sowing; throughout life for crops harvested during the summer months; a pre-harvest application will aid lifting when the soil is dry.

Treatment—If the soil is dry enough, before sowing for crops drilled between April and June inclusive, give 1 in. in April and up to 2 in. in May and June. Outside those months the soil is usually moist enough so that watering is not necessary. For crops harvested during the summer months between April and August, give 1 in. at 1 in. S.M.D. If the soil is dry at harvest, give 1 in. to aid lifting.

Parsnips

Prospect—Marginally economic.

Period—Pre-sowing only for crops drilled after mid-April.

Treatment—If the soil is dry, give 1 in.

Peas, (Green, Vining and Harvesting Dry)

Prospect—Routine treatment for all crops except those grown for picking before the end of June. The value of these early crops depends primarily on date of picking; as irrigation may delay maturity, do not water.

Period—At early flowering and at pod-swelling. Yields will not be increased by irrigation at other times. Watering before early flowering usually increases haulm growth, which is not required. Watering after flowering can increase number and size of peas.

Treatment—If the S.M.D. is 1 in. or more at either of these stages of growth of the crop, give 1 in.

Potatoes, Early and Canning

Prospect—First priority.

Period—May to June for early potatoes. Emergence to lifting for canning potatoes.

Treatment—Apply 1 in. (2·5 cm) whenever the S.M.D. reaches 1 in. (2·5 cm).

Potatoes, Second Early

Prospect—Routine treatment.

Period—May to July.

Treatment—From the time tubers reach $\frac{1}{4}$ in. (0·6 cm) diameter apply 1 in. (2·5 cm) at S.M.D. of 1 in. (2·5 cm) on Class A soils and $1\frac{1}{4}$ to $1\frac{1}{2}$ in. (3·2 to 3·8 cm) whenever the S.M.D. reaches this level on Class B soils.

Potatoes, Maincrop

Prospect—Routine treatment.

Period—June to August.

Treatment—From the time tubers reach 1 to 2 cm diameter apply 1 in. (2·5 cm) at S.M.D. of 1 in. (2·5 cm) on Class A soils and $1\frac{1}{2}$ in. (3·8 cm) at S.M.D. $1\frac{1}{2}$ in. (3·8 cm) on Class B and C soils.

Irrigation can also be used to control common scab. Tubers are susceptible as they form and wet soil conditions at this stage prevents the development of the disease. If the S.M.D. is over $\frac{1}{2}$ in. (1·3 cm) when the plants cover 10 to 20 per cent of the soil surface, return the soil to field capacity. During the following *four-week* period S.M.D. should not exceed $\frac{3}{4}$ in. (1·8 cm) if a high proportion of scab-free tubers are to be produced. Wet soil conditions at tuber formation also encourages tuber set, and caution should be exercised using this technique with varieties which naturally have high tuber numbers.

Radish

Prospect—First priority.

Period—Throughout life, from April to August.

Treatment—Give 1 in. at 1 in. S.M.D.

Rhubarb in the Open

Prospect—Routine treatment.

Period—After pulling has stopped.

Treatment—On Class A and B soils if the S.M.D. is more than 2 in. give $1\frac{1}{2}$ in., on Class C soils if S.M.D. is more than 3 in. give 2 in.

Spinach

Prospect—Routine treatment for summer crops.

Period—Pre-sowing; throughout life until August.

Treatment—If the soil is dry enough before drilling, give 1 in. in April, up to 2 in. thereafter. Between May and August irrigate the growing crop with 1 in. at 1 in. S.M.D.

Sugar Beet

Prospect—Marginally economic.

Period—July and August.

Treatment—On Class A soils apply 2 in. (5·1 cm) whenever the S.M.D. reaches 2 in. (5·1 cm) after the leaves meet between the rows. On Class B soils apply 2 in. (5·1 cm) whenever the S.M.D. reaches 3 to 4 in. (7·6 cm to 10·2 cm). Class C soils require irrigation only in exceptionally dry seasons.

A period of 3 to 4 weeks should be allowed between the last irrigation and harvest otherwise sugar percentage may be reduced.

Swedes (for Vegetable Market)

Prospect—Marginally economic.

Period—Pre-sowing, May and June.

Treatment—If the soil is dry enough at drilling, give up to 2 in.

Turnips (for Vegetable Market)

Period—Pre-sowing, from beginning of root swelling onwards.

Treatment—If the soil is dry enough at drilling, give 1 in. in April or up to 2 in. in May. From the beginning of root swelling until harvest or the end of July, give 1 in. at 1 in. S.M.D.

NURSERY STOCK AND FLOWERS

There is little experimental data on which to base recommendations. The guidance offered is based mainly on work carried out on allied crops or on practical experience.

All Crops

Adequate soil moisture must be ensured for sowing or planting. Irrigation is best applied during the preparation of the land and should be sufficient to give adequate moisture throughout the soil profile.

Water transplanted crops to aid establishment whenever necessary, including dormant plants which are planted later than normal from a cold store. Continue irrigation until establishment is complete.

Established Plants—Annual and Biennial Flower Crops

Prospect—Routine treatment.

Time—Throughout life. Do not water overhead when plants are in flower.

Treatment—On all soils apply 1 in. at 1 in. S.M.D. With some subjects irrigation may delay flowering and may result in excessive leaf.

Chrysanthemums Grown in the Open

Prospect—First priority.

Time—Throughout life. Do not water overhead when plants are in flower.

Treatment—On all soils apply 1 in. at 1 in. S.M.D.

Gladioli

Prospect—Routine treatment.

Time—From planting to when leaves are 1 ft high.

Treatment—On all soils apply 1 in. at 1 in. S.M.D.

Other Bulbs, Corms and Tubers for Flower or for Bulbs

Irrigation is seldom necessary for these crops.

Other Herbaceous Perennials

Prospect—Routine treatment.

Time—Throughout life.

Treatment—On Class A soils apply 1 in. at 1 in. S.M.D.; on Class B and C soils apply 2 in. at 3 in. S.M.D.

Biennial Bedding Plants

Irrigation prior to lifting aids lifting and establishment on replanting.

Trees and Shrubs

Prospect—Routine treatment.

Time—May to July inclusive.

Treatment—On Class A soils apply 1 in. at 1 in. S.M.D.; on Class B soils apply 1 in. at 2 in. S.M.D.; on Class C soils apply 2 in. at 3 in. S.M.D.

Time—Just before lifting (dry soil conditions only).

Treatment—Apply about 1 in. of water.

Container-grown plants need continued and frequent irrigation throughout their life. This is applied through spraylines or by automatic systems such as trickle systems or capillary or sand benches as described on pages 66–72.

FRUIT CROPS

With tree and bush and cane fruits, during the growing season the fruit crop is developed and matured and the future fruit-bearing wood or cane is also produced. Soil moisture deficiency during the growing season is a hazard not only to the current season's crop of fruit but to future crops as well.

As with other crops the moisture requirement of fruit is high when the weather is hot and bright, and low when it is cool and dull. It is therefore difficult, if not impossible, to define in precise terms a range of soil conditions during which adequate moisture for crop needs may be assumed. The deeper the soil and the more extensive and fibrous the root system the better buffered are fruit plants against change in weather; for this reason mature trees are less affected than young trees, and trees generally are affected less by hot dry spells than are soft crops such as bush and cane fruits.

Fruit Trees

Poor growth, resulting in low yield is probably the most general indication that summer rainfall and soil moisture reserves together may be inadequate to supply the full needs of an orchard. A succession of favourable seasons, with high summer rainfall, may permit a more rapid increase in tree size and cropping capacity of trees on relatively shallow soils, but may well render them more susceptible to the drastic effects of a dry season, because their roots have fully occupied the soil and yet lack depth.

Because the amount and distribution of summer rainfall is so erratic the need for extra water in English fruit orchards will be determined, more often than not, by the depth and texture of soil upon which the roots depend for their moisture. Provided there are no physical barriers to the roots, the soil and subsoil, if sufficiently deep, can provide a very large part of the water need of mature fruit trees during a summer season.

In the Midlands, East and South-east of England fruit trees need to draw 4 to 6 in. of water from the soil moisture reserves in most years but in an exceptional year, like 1959, double these reserves may be required. If orchard soils cannot supply these amounts then supplementary watering should form part of the normal orchard management.

Adequate soil moisture in May and June greatly assists the production of new shoots while water applied later in the season maintains foliage in an effective condition and guards against reduced fruit size; this effect on fruit size is most pronounced in summers of prolonged drought. Yields increase with successive seasons of irrigation and the benefits in relation to fruit size, therefore, decrease unless a system of pruning or fruit thinning is carried out to prevent too heavy a crop load. Where large increases of crop have occurred with irrigation they have generally been achieved without gain or loss in fruit size compared with crops harvested from unwatered trees.

The real worth of irrigation is measured by the cumulative effect on cropping. A delayed cropping response is not unusual in fruit trees and is due to the time taken for irrigation to increase the shoot to root ratio and for fruit buds to develop and crop on the new growth.

Obviously, a progressive increase in cropping cannot continue indefinitely and in the absence of any other limitation to growth it can be expected to continue only to the point of full occupation of the aerial volume of the tree.

When water for irrigation is limited, it is more economic to irrigate less frequently than is ideal on a larger acreage than to concentrate the maximum amount of water on a smaller acreage.

No deterioration has been found in the keeping quality of Laxton's Superb apples from heavily irrigated trees and no marked effects of the different levels of irrigation were found on Cox's Orange Pippin apples. Bitter pit may be reduced where irrigation aids calcium intake and retention in fruit, but may be increased where irrigation results in larger fruit and, in consequence, dilutes the calcium content of the fruit.

Management and manuring of fruit trees may have to be modified to suit the change in growth, vigour and cropping capacity that occurs when irrigation is practised regularly. Smaller applications of nitrogenous fertilizer and less frequent mowing in grassed orchards will reduce any tendency towards an over-nitrogenous condition of the trees.

An irrigation guide for all fruit crops is provided in Table 11, but where irrigation facilities are very limited by water or equipment the aim should be to apply one or two timely irrigations during the season rather than to wait until a serious deficit has developed. It is more economical to apply larger quantities of water infrequently rather than the reverse, unless the quantity of water available prevents a heavy overall application of say 2 in. The maximum possible pentration of the water should be ensured. This can be done by concentrating the application to a limited area around each tree or along each tree row, leaving the mid-alley space unwatered.

The effects of irrigation on cropping that have been described can only occur if the irrigation is regular so that the fruit crops may first adapt to the new conditions and then provide the crop advantage that the adaptation affords.

Bush and Cane Fruits

Black currants and raspberries have been found to respond well to irrigation even on deep soil of good water holding capacity. The rapid development of the small fruits and the concurrent development of new wood to carry the following season's crop demand a readily available supply of water to avoid checks to either process during a spell of hot dry weather. Fruit quality and quantity appear to be equally benefited by timely irrigation.

BLACK CURRANTS

As with apples the cumulative effect of irrigation on yield can become very large. Irrigation increases black currant crops partly by increasing fruit size but mainly by increasing bush growth; it is important, therefore, to irrigate to maintain growth during the entire pre-picking period from May to August. Exceptionally, some irrigation may be needed in April if the season is early and April is rainless. The growth of black currant bushes begins to diminish when day length shortens so that there is seldom need for watering after early August.

Black currant roots tend to be concentrated in the row over a width of about 4 ft and maximum drying occurs in this zone to a depth of about 18 in. In many plantations, therefore, water could be used more economically by irrigating row widths of 4 ft rather than the whole area.

It is important to maintain adequate soil moisture especially during hot weather when most adverse effects are likely to occur. Irrigation with 2 in. of water is recommended when the soil moisture deficit (based on potential transpiration loss for full cover less rainfall) exceeds this amount. A 3 in. deficit will probably be tolerated in the better loams as is indicated in the Irrigation Guide in Table 11.

RASPBERRIES

The Irrigation Guide, Table 11 gives the period for raspberry irrigation as June to July. Raspberries are as sensitive to soil moisture status as black currants but there is often no advantage in irrigating to stimulate growth because of the common practice of reducing fruiting canes to a given density and height by thinning and tipping each year. If the management system used can accommodate the growth vigour stimulated in raspberries by regular watering, without overcrowding in the rows, or in the first year after planting, it would be advantageous to follow the recommendations for black currants.

Increased crop weight from irrigated canes in continuous rows is due mainly to an improvement in berry size. This can often be achieved by a single irrigation of 2 in. of water when the drupes begin to swell or just as the berries show a pink tinge. Over a period of 6 years, watered raspberries (varieties Malling Jewel and Malling Exploit) yielded a seasonal average of 184 lb of fruit per 100 ft of row compared with 145 lb from unwatered canes. The increased yield was remarkably consistent after the first year.

Raspberries are sensitive to soil compaction particularly under wet conditions; it is therefore advisable to avoid irrigation during picking.

Strawberries

The Irrigation Guide, Table 11, gives several response periods for strawberries, and watering should be carried out at the right stage in the plants' development to achieve the required response.

For fruit size, soil water supply should be plentiful during the period preceeding picking. Strawberries have an early cropping season so that high soil moisture deficits during the fruit swelling period are less likely then with other fruit crops. Strawberries are, however, relatively shallow-rooted and fruit swelling may be adversely affected by the weather giving rise to high transpiration rates even when deficits are relatively low. Pre-blossom watering favours foliar development with no effect or (perhaps even an adverse effect) on yield; post-blossom irrigation is chiefly effective for maintaining fruit swelling.

The heavy crops produced on mutally competing plants in matted rows are particularly susceptible to drought effect. It is therefore advisable to apply 1 to 2 in. of water to matted rows whenever the calculated soil moisture deficit has reached 2 in. during the period from full flower to completion of fruit swelling. Non-competing spaced plants can withstand longer periods of drying without adverse effect on yield; even so, results of experiments have shown highly economical increases in marketable fruit from single applications of between 1 to 2 in. of water given during the fruit swelling period

For runner production water supply should be plentiful during the whole period of production and light and frequent irrigations would seem to be of greatest advantage.

For runner establishment, water supply should be plentiful at planting; the guide suggests 1 to 2 in. of water applied before planting but at least 24 hr should be allowed between irrigation and the start of planting.

HOPS

The irrigation of hops can be of benefit on shallow soils or in very dry areas of the country. When setting out mist-propagated plants or strap cuttings in the nursery, or late spring planted bedded setts in a hop garden, irrigation can be of great value in their establishment particularly if May and June are dry.

An established hop plant carries a relatively small amount of top growth in May and June, and the large permanent root system is normally able to supply moisture needs during this period.

In dry summers the irrigation of established hop gardens is most likely to be of benefit in June and July. At this time and during early August the above ground parts of the plants are at their maximum stage of growth and demand for water and nutrients is likely to be at its peak.

Where a grass sward is used in hop gardens the competition for moisture may increase the need for irrigation. The use of water by the sward will depend on its composition. If it is shallow rooting and mainly composed of annual meadow grass it will tend to die off during dry weather and thus compete less with the crop for moisture.

Where irrigation is applied to hops, diseases such as verticillium wilt and downy mildew may be encouraged to spread.

Table 11

Summary of irrigation treatments for fruit and hops

Crop	Response period
Strawberries, runners at planting*	Date of planting to September
Strawberries, fruiting	May to June and August to September
Strawberries for runner production	May to July
Plums, cherries and gooseberries	May to July
Hops	May to July
Hops, setts*	May to August
Apples and black currants	May to August
Pears	May to August
Raspberries	June to July
Loganberries and red and white currants	June to July
Nuts and blackberries	June to August

* The ground for strawberry runners and hops setts planted in summer will need 1 to 2 in. water before planting; the S.M.D. should be accumulated from this irrigation date. The S.M.D. should be accumulated from 31 July, ignoring the S.M.D. at this date.

Rooting depth	Crop	Irrigation plan for soil types		
		A	B	C
		Inches available water per foot		
		less than $1\frac{1}{2}$ in.	$1\frac{1}{2}$ in. to $2\frac{1}{2}$ in.	$2\frac{1}{2}$ in. or more
Less than 18 in. or very gravelly soil of any depth	All fruit crops and hops	1 in. at 1 in. S.M.D.	1 in. at 1 in. S.M.D.	2 in. at 2 in. S.M.D.
18 in. to 24 in.	All fruit crops and hops	1 in. at 1 in. S.M.D.	2 in. at 2 in. S.M.D.	2 in. at 3 in. S.M.D.
24 in. to 48 in.	All fruit crops and hops except mature fruit and nut trees	2 in. at 2 in. S.M.D.	2 in. at 3 in. S.M.D.	2 in. at 3 in. S.M.D.
24 in. to 48 in.	Mature fruit and nut trees	2 in. at 3 in. S.M.D.	2 in. at 3 in. S.M.D.	3 in. at 5 in. S.M.D.
More than 48 in.	All fruit crops and hops except mature fruit and nut trees	2 in. at 2 in. S.M.D.	2 in. at 3 in. S.M.D.	2 in. at 3 in. S.M.D.
More than 48 in.	Mature fruit and nut trees	2 in. at 3 in. S.M.D.	3 in. at 5 in. S.M.D.	No irrigation

GRASSLAND

Prospect—Over a run of years, full irrigation is likely to increase grass yields by about 30 per cent, varying from dry years when irrigation may double herbage production, to wet years when little irrigation is needed and little response is obtained. An important benefit from irrigation is that herbage production is more uniform and more reliable both from season to season and within each season, though it cannot eliminate the reduction in growth which follows the flowering period. Another advantage is that mid-season establishment of grassland can be practised with greater confidence.

Grassland irrigation is unlikely to be profitable for stock-rearing or for traditional beef and sheep enterprises. Its most suitable place is on the intensive dairy farm. Even here, the resulting increased herbage production is unlikely on its own to support sufficient extra output to repay the cost of irrigation except where installation costs are extremely low. In practice the profitability of irrigation often depends on the extent to which the resulting more certain and uniform growth of swards allows a general increase in stocking rate, a reserve of grass no longer being needed to meet the risk of drought. This is much affected by the circumstances of each individual farm and farmer. At one extreme is the farm, already heavily stocked, which copes with a dry season by buying in feed or by keeping a reserve of hay or silage. Irrigation will show relatively little return here. At the opposite end of the scale is the farmer who is particularly concerned if grass is scarce and who keeps his stocking rate down to a level appropriate for the drier season. The introduction of irrigation may allow this farmer to double his stock-carrying capacity.

The profitability of irrigation is also affected by the way it is exploited. If cow numbers can be increased on the existing grass acreage with little extra expense on buildings and labour, this will normally be the more profitable method. If herd size remains unaltered and the grass acreage is reduced, then the cost of irrigating the remaining grass must be offset against the nett value of the alternative cash crop grown on the released acreage.

Period—Throughout life. Because established grass covers the ground throughout the season, it has a greater annual water requirement than other crops. In many parts of the south, midlands and eastern parts of England, irrigation need averages 5 in. per year and can be more than double this in dry years.

Treatment—Despite the ability of many grasses to remove water from depths of 2 to 3 ft the growth of grass swards is reduced once the soil moisture deficit exceeds about 1 in. and is substantially reduced at a 2 in. deficit. This sensitivity to water shortage arises because the surface horizons of the soil are the first to be depleted of water and this reduces or prevents uptake of nitrogen which is largely confined to these horizons. The first effect of drought is therefore, at least in part, due to lack of nitrogen rather than lack of water.

In intensive irrigation schemes, grazed fields should be irrigated when the soil water deficit reaches 1 to $1\frac{1}{2}$ in. A deficit of $1\frac{1}{2}$ to 2 in. can be allowed in fields put up for conservation. These recommendations apply regardless of soil type. Light and medium soils should be irrigated to field capacity but a small deficit of $\frac{1}{2}$ in. can be left on heavier soils to allow for any subsequent rainfall.

Irrigation at less than the full amount may sometimes be appropriate if water is scarce or the equipment is primarily intended for other crops. Until a soil water deficit of about 4 in. is reached, ½ in. of irrigation after each nitrogen application will give a larger response *per acre inch* than full irrigation, but at the expense of a lower response per acre. Such a policy must be very carefully controlled and is no justification for getting behind with a full irrigation schedule.

GLASSHOUSE AND PROTECTED CROPS

Irrigation in Glasshouses and Structures

The water relationships of crops grown in glasshouses are different in many ways from those of outdoor crops. In a bulletin such as this the subject cannot be dealt with fully.

In glasshouses there is no natural rainfall and soil moisture reserves can be replenished only by artificial means. Furthermore, the glasshouses structure so alters the environment that crop water loss is largely determined by the incident radiation and in practice this is the only factor that needs to be considered.

EFFECT OF MOISTURE REGIME

When a crop is grown in soil, water is extracted by the roots and the water content of the soil is progressively reduced until further water is added to make good the depletion. When the moisture content of the soil is allowed to fluctuate only between field capacity and a predetermined state of dryness, a crop is subject to a specific 'moisture regime'. When the fluctuation is small, the regime is said to be 'wet', and where the fluctuation is large, the regime is said to be 'dry'.

The moisture regime maintained in a soil markedly affects the growth, yield and quality of a crop. Much more work needs to be done on a wider range of crops, but the following general principles have been suggested by past experience.

Where plants are set out into the soil at or near field capacity, growth is not affected by added water so long as the roots are growing strongly into moist soil. Once the soil volume has been thoroughly explored by developing roots, the crop becomes responsive to changes in moisture regime, maximum growth being obtained when the soil is maintained near to field capacity.

During the early period of growth immediately after planting, the crop needs little or no added water. Indeed, watering may have an adverse effect. With early tomatoes provided they are planted when 50 per cent of them have one flower open on the first truss, a settling-in watering of about 1 pint per plant should be given. Routine watering with liquid feeds should not begin until overhead damping, applied at least once daily, is insufficient to keep the plants turgid. This is usually within two weeks after planting when the earliest fruits are setting. The first watering should bring the soil back to field capacity in the root zone. Watering is then repeated two or three times weekly, and should be based on one of the estimation methods described later, for maximum growth and yield are obtained where the soil is never allowed to fall much below field capacity.

With short-term crops such as winter lettuce, it is possible to grow the crop to maturity without watering after planting provided the soil is deep enough and has sufficient water-holding capacity to maintain root development.

Moisture regime also influences root distribution, vegetative growth and fruit quality in the tomato. Wet regimes result in maximum root development in the top 4 in. of soil, whereas dry regimes give maximum development in the 8 in. to 12 in. zone. This distribution of the root system emphasizes the importance of subjecting plants to a consistent irrigation treatment throughout their life. A plant which develops most of its roots in the surface soil early in the season as a result of a wet regime will be badly checked if the soil is suddenly dried out later in the season. Similarly, a plant which develops the majority of its roots at lower levels will be checked if water fails to penetrate to these levels later. Once a particular watering procedure has been established, it is therefore unwise to alter it, although it has been shown that where drip or trickle irrigation systems are used, occasional flooding of tomato borders with sufficient water, applied through hoses or overhead spraylines, to pass into drainage can encourage vegetative growth and improve fruit set and development. This is probably the result of correcting localized areas of dry soil and high salt concentration, and providing a moister and cooler glasshouse atmosphere. Dry regimes result in lower yields of better quality and more uniformly coloured fruit than wet regimes, although water regime is not the major factor involved. Vegetative growth is affected by moisture regime and fruit load, and the effects of both factors are additive, weakest growth occurring with a dry regime and heavy fruit load.

PRINCIPLES OF WATER LOSS

By growing plants on weighing machines and in lysimeters it has been shown that water loss depends on solar radiation, and that the proportion of this used in evaporating water from a tomato crop increases as the plants grow and intercept more radiation. This proportion, called the 'water factor', and its relation to plant height is shown in Fig. 22. The water factor increases

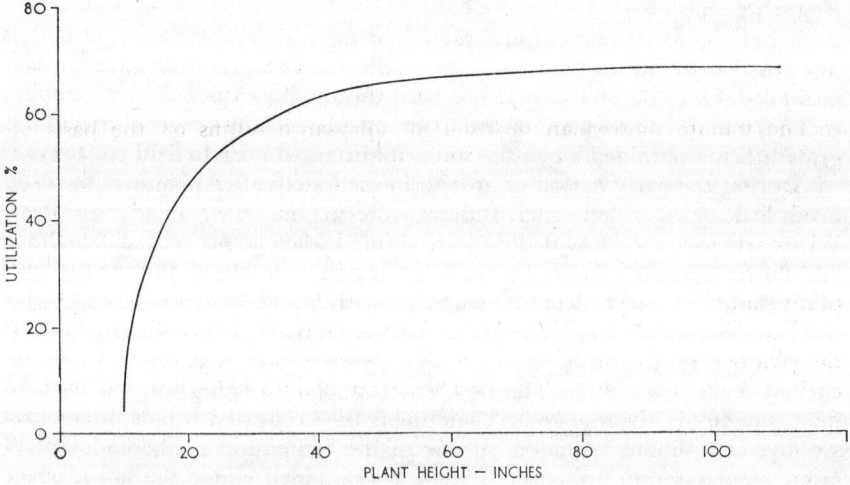

Fig. 22. Proportion of solar radiation utilized in relation to plant height

steadily during the early stages of plant growth, but becomes constant at about 70 per cent, when a full leaf cover is reached at a height of 4 to 5 ft.

Recent research has suggested that a 70 per cent factor is generous in respect of the majority of tomato plants in a 30 ft wide, east-west vinery house. It is most nearly applicable to those along its southern side, which may have a mean transpiration rate through the summer about 10 per cent greater than those through the centre of the house. Excess irrigation will go into drainage under the conditions specified.

Radiation is recorded by solarimeter at a number of centres, and its conversion by calculation into a tomato 'water figure', issued by A.D.A.S., has led to a useful practical application in surrounding areas of these principles. There is no reason to doubt that, with some modification, it can also be used with other glasshouse crops.

USE OF THE TOMATO WATER FIGURE

The water figure given in gallons per acre of glasshouse is used as an irrigation guide by growers in areas near to one of the research or experimental horticulture stations which provide the calculations. It is available to growers on request. It can be used without adjustment if healthy plants, which have attained full ground cover, are growing in well-drained soil in a modern, single-span house with clean glass. In conditions different from these or in more remote areas the local A.D.A.S. Horticultural Adviser should be consulted.

For plants which have not attained full ground cover the following adjustments to the issued figure should be used:

Plant height	10 in.	15 in.	20 in.	40 in.	60 in.+
Percentage of water figure	15	50	70	90	100

The soil must be at field capacity when calculated watering begins, for the figure given is intended to maintain field capacity in the glasshouse borders.

The water figure has a limitation in that it is necessarily issued in arrears. Irrigation which has been applied can then be corrected if a deficiency needs to be made good. A mean water figure, based on several years' figures for a particular week can provide an advance guide.

The tomato figure can be used for chrysanthemums on the basis of glasshouse area, adjusted for ground cover as follows:

Plant height	6 in.	9 in.	12 in.+
Percentage of water figure	86	93	100

It can also be used for carnations on the basis of total bed area, adjusted as follows:

Plant height	6 in.	9 in.	12 in.	15 in.	21 in.	33 in.+
Percentage of water figure	71	86	93	100	110	142

This calculation has been used successfully for several years. Increasing the tomato figure to 142 per cent counteracts drying out at the edges of the beds. If an irrigation figure per area of *glasshouse* is required, the above percentages must be reduced by the proportion of the *bed* area. If this is 70 per cent, then the full ground cover water requirement per acre of glasshouse equals the unmodified tomato figure.

Whilst a calculated irrigation method may bring simplicity, it should be co-ordinated with experience based on the appearance of the crop and the use of an auger to assess soil moisture.

Protected Cropping

The term 'protected cropping' refers to the growing of crops under temporary, portable or movable equipment such as lights, cloches, frames and tunnels. Such equipment is usually constructed mainly of glass but many forms of transparent plastic materials are also used. The protection shields the crop from the wind and tends to increase the temperature especially in the soil, but in general it prevents the rain from reaching the ground immediately surrounding the crop. Some forms of cover such as Dutch lights or glasshouses exclude almost all natural rainfall, others such as cloches or other types of frames admit part of the rain. It therefore follows that the soil moisture conditions of crops under low protective cover deserve special consideration.

The protection may be provided for a short period of only a few weeks when the limiting soil moisture deficits are not attained and therefore no irrigation is needed. Longer periods of protection amounting to 3 or 4 months often comprise a winter programme when with short days and weak sunshine growth is slow and water use is relatively small. Protection during late spring or early autumn presents greater soil moisture problems and any form of cover during summer demands the use of irrigation to obtain the best results.

Plants under glass or plastic-covered structures which provide conditions comparable with those within permanent glasshouses have evaporation rates which are also comparable and the water figure (see page 93) can be used as a guide provided the glass or plastic is clean. In summer, when ventilation has to be freely given in plastic-covered structures, air movement may raise the irrigation need above the water figure. Judgment must not be distorted by condensation on the underside of the cover which is often visible in the early morning. This moisture has been distilled from the soil and, on re-evaporation, it will be carried away by the air which is constantly moving past it. Few structures are airtight and even the best constructed glasshouses have several air changes per hour. In autumn, winter and spring, when ventilation may be very limited in plastic structures, the irrigation need will probably be less than the water figure.

It is essential that the soil should be brought to a moisture state near capacity before it is sown or planted and covered. This may be accomplished by rainfall alone but any deficit resulting from dry weather or a previous crop must be eliminated by irrigation; sprinklers are generally used for this. Before sowing or planting the land should have started to dry out on the surface.

When the soil is thus brought to capacity at the start of the protected period there will often be enough moisture within the root range to finish the crop without check or to last until the protection is removed. This is particularly true for overwintered crops or those covered for a short period in autumn or early spring. It is not true in general for summer crops.

EXAMPLES OF WATER NEEDS
Overwintered lettuce seedlings in frames

All techniques are directed towards the preparation of a firm seedbed with the maximum supply of moisture in the topmost layers of the soil to sustain the seedlings throughout the winter. The preceding crops are generously manured and the residues are well incorporated in the top soil. The site carries no crops for about eight weeks prior to the October sowings, and during this time it is allowed to receive all the natural rainfall and to compact itself firmly in a way difficult to bring about by direct mechanical consolidation. If late September has been dry it may be necessary to add about $\frac{1}{2}$ in. (2 gal per sq. yd.) of water by sprinkler or sprayline prior to the final preparation for sowing.

Correctly prepared in this manner, the seedbed will not require any further watering until the seedlings are moved later in winter or early spring. Even if the transplanting takes place as late as March, little irrigation is required.

Lettuce for early winter cropping

Lettuce sown in early August and planted under glass cover will often come to maturity in November or early December without additional irrigation provided that, as before, the soil was at field capacity at the start. This is most likely to be the case where cloches of small span or glass frames are used which allow some entrance of rain and at the sides of which there is some seepage through the soil of the water which has fallen outside the protective cover. A similar crop sown under lights will probably need supplementary water. This is best given when the crop is half-grown in October to the extent of 2 to 4 gal per sq. yd.

Lettuce for April cropping

This crop is sown in October and planted in November. Provided that the soil is correctly prepared the critical period for soil moisture will not occur before March. If the weather then is sunny, if the soil is light and retains little moisture, and particularly if the crop is under lights, an irrigation of 2 to 4 gal per sq. yd. may be needed. Again, this should be applied when the crop is half-grown and the best results will not be obtained by later watering.

Spring and summer cropping

From March onwards the strength of the sun increases considerably and the transpiration of crops increases proportionately. Soil moisture reserves can soon become exhausted and additional water is likely to be needed for all crops under glass. As a guide to the amounts of irrigation to be applied, the weekly water figure or the potential transpiration figures for crops in the open with no allowance for natural rainfall can be followed.

General considerations

Rainwater or irrigation applied over the top of protecting cover tends to collect along the sides, although some flat-topped frames allow penetration within the covered area. The water which has collected along the sides will tend to seep sideways slowly through the soil, but this seepage will only take place to any appreciable extent when the soil is moist. If the soil under the cover is already dry the sideways movement is slow and very restricted in extent. With wide types of cloche or frame this seepage is of very little value to the crop. If the width is less than 2 ft, some benefit is obtained and some of the roots of the crop will in fact have grown beyond the limits of the cover and will be drawing directly on the moisture in the soil which has fallen as rain outside. When preparing land for cloches it is sometimes possible to arrange shallow drills so that the water falling outside can run inside the cover. Devices exist for garden use with the same object in mind. The main difficulty is that during the winter months it is desirable to keep rain off the crop and the soil surrounding it. Water is only required in limited amounts on few occasions. During the greater part of their use therefore, cloches should exclude rain and not admit it.

The economic advantages to be gained by the use of protected cropping methods are limited by labour costs. Since summer labour costs can chiefly be incurred by the need to irrigate, methods of irrigating under low cover must be designed with this consideration in mind.

Trickle irrigation can be used for some protected crops and so can perforated polythene tubes. Both methods help to reduce labour costs. Double span ranges of lights may be fitted with irrigation line in the ridge, but the water is effectively applied only with very low crops which permit the *even* uninterrupted spray of water. Internal ridge irrigation equipment may have diffuser nozzles with alternate throws in opposite directions, or 4-cut nozzles giving a circular throw.

Crops which have been started under protective cover and which are finishing with the cover removed may of course be irrigated in the same manner as for an open ground crop.

Crops grown under glass cover can be protected against frost by the use of continuous sprinkling throughout the dangerous night period (see page 97 for further details of frost protection by sprinkling).

FORCED RHUBARB

Immediately the shed has been filled the roots should be watered with a powerful flow from a hose to wash all the soil from the crowns into the spaces between the roots. If the buds are left covered with soil the leaf stalk bases will become soiled and the quality impaired.

After the initial watering it is usual to water rhubarb about once a week at the rate of about 1 gal of water per sq. yd of shed. During the pulling period such watering should be done immediately after pulling. Overhead sprinkler irrigation systems are used for this purpose.

MUSHROOMS

Routine watering of mushroom crops is still by hose pipe and hand lance through various rose types. The most popular rose type is the oval headed fine spray which can be governed to apply a gentle fall of water on to the cropping beds. Other types, such as roundhead roses and watering rods varying from 10 to 21 in. in length have been used in special circumstances. Watering rods, for example, are fairly easy to move between tightly packed trays and give an evenly fine gentle spray to the crop. The flow of water is governed by regulating the pressure at the tap or through a trigger hand lance.

Automatic watering systems have been devised but they are not employed widely on commercial farms. A completely automatic system, developed at the Glasshouse Crops Research Institute, using nylon felt wicks across the beds being fed from water troughs, has been tried. This form of capillary watering has problems of commercial application with the present arrangements of trays in growing houses and the effect of heavy flushes disturbing the wicks. It is used for research and experimental work where constant moisture levels are required. Where shelf beds are used, some being 4 or 5 ft high in a growing house, movable watering racks have been devised. In Holland this technique finds more favour as there are a greater proportion of shelf growers than in this country. The racks are on wheels and the watering lance can be adjusted to the shelf height, the racks being then pushed up and down the pathways. The system is mainly suitable for high beds and obviates the need to use steps.

Calculated water application is not yet accepted practice with mushrooms and it is up to the grower's judgment when to apply water and how much is required. This is an extremely important part of mushroom culture that affects yield, quality and disease control. Unwise application can lead to lack of pin head initiation or damage from bacterial disease. The main technique is based on bringing the crop in with light spraying over, increasing as the pin heads develop with heavy applications between flushes of mushrooms. This technique is more simple during the winter months when evaporation from the beds is rapid due to warm air used for heating. In the summer when humidities are high, it is difficult to assess the water required and there may be a tendency to over-water.

FROST PROTECTION

Of the various methods available for preventing frost damage to outdoor crops only those using irrigation equipment are dealt with in this bulletin. In recent years irrigation equipment has been increasingly used for frost protection purposes. The two methods of using water in this way are to pre-wet the soil as a preventative measure or to sprinkle continuously throughout the period when frost is actually occurring.

Pre-Wetting or Advance Irrigation

This method aims at increasing the effectiveness of the soil as a reservoir of heat, by applying irrigation not long before a possible night frost. Wet and compact soil is a better conductor of heat than dry and loose soil. Irrigation both wets and compacts the soil thus leading to a greater total heat absorption on a warm day and a greater upward transfer of heat from the soil to the air during a clear cold night. Experiments have shown that the air 1 ft above wet bare soil, on a still frosty night, can be about 1°C warmer than that above dry bare soil.

The technique is of value only for low-growing crops and against slight radiation frosts. It has been used mainly for the protection of early potatoes and strawberries.

An application of $\frac{1}{3}$ in. of water every four days is recommended and, on this basis, equipment covering 1 acre can be moved over some 20 acres in the four-day-period between applications. The technique enables a relatively small amount of equipment and water to be used over a large area.

Continuous Overhead Sprinkling

Continuous water sprinkler systems can be effectively used against radiation frosts but are unlikely to give complete protection from wind frost as an excessively high rate of water application is required, and the wind can also severely distort sprinkler pattern. Increased damage can result from starting sprinkling under severe wind frost conditions. Continuous sprinkling is used commercially for the protection of the blossom of fruit crops on what is still a comparatively small scale in Britain, but is more extensively practised in Italy, Germany and the U.S.A. It can also be used to protect vegetables and flowers providing they can bear the ice load and for protection of crops under Dutch lights or under polythene tunnels. The installation whether permanent or temporary must be sufficient to cover the entire area to be protected.

The protective effect depends on the transfer of heat from the added water, mainly by the release of latent heat on freezing of the water as it turns to ice. This provides approximately 335 kilojoule/kilogramme (80 calories/gramme) of water frozen which is equal to about 1,440,000 B.T.U. per 1,000 gal. In order to achieve protection, a film of water must be maintained on the plant throughout the period of frost. As long as there is free water available to freeze, the temperature of the ice and water mixture and the plant will remain near to 0°C. At this temperature blossoms are unharmed as the critical temperature for open blossom is between $-2 \cdot 8°$ and $-2 \cdot 2°$C. If sprinkling is interrupted for more than a few minutes once the bushes or trees are wet, all the water may freeze and the flowers are then likely to sustain *more* damage than if no sprinkling had been done. Attention must be given to ensure adequate land drainage and the soil should be able to accept the sprinkling rate without suffering structural damage or becoming waterlogged.

SPRINKLER TYPE AND SPACING

Choice of correct type of sprinkler and the sprinkler spacing and operating pressure are most important. The spring assembly on each sprinkler should be correctly tensioned to give one rotation every 30 to 60 sec in order to achieve adequate rewetting of the crop; it should also be enclosed to prevent the moving parts from icing up. The lack of balance associated with single nozzle sprinklers used for frost sprinkling requires the standpipes to be firmly supported to prevent whip, which can affect the speed of rotation and range of the sprinklers.

An operating pressure in the region of 50 to 60 lb per sq. in. at the sprinkler nozzle is required with the present types of rotary sprinkler to achieve adequate range and to break up the droplets.

IRRIGATION IN PRACTICE

A triangular spacing of sprinklers is usually adopted, in order to obtain as uniform an application of water as possible. To reduce the costs of installation, there has been a tendency to increase sprinkler spacings, and at such extended spacings an equilateral triangle arrangement will usually ensure a better distribution and coverage of water than the more usual isosceles triangle arrangement. For example, the equilateral arrangement of 80 ft between sprinklers on the line × 69 ft (say 70 ft for convenient pipe length) between lines requires 7·8 sprinklers per acre, compared with 6·8 sprinklers per acre used in the isosceles triangle arrangement of 80 ft × 80 ft. An 80 ft spacing is taken for the example; actual spacings will depend on the sprinkler type and nozzle size and tree planting distance, but the equilateral principle still applies.

Equilateral triangle arrangement of sprinklers:-

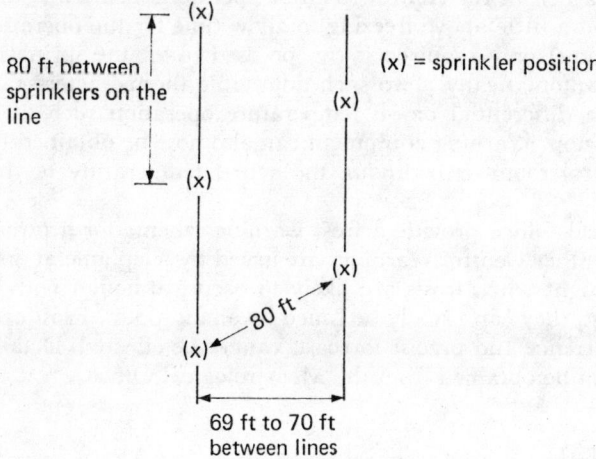

Isosceles triangle arrangement of sprinklers:-

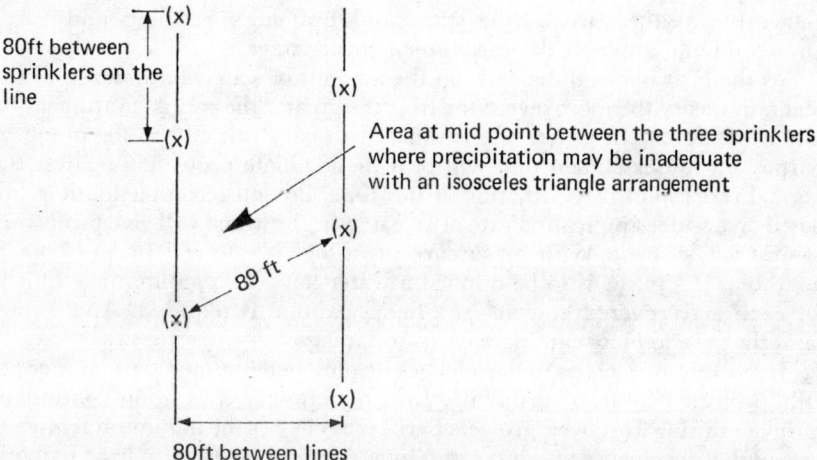

Fig. 23. Equilateral/isosceles triangle arrangement of sprinklers

OPERATING TEMPERATURE

The sensing thermometer should be unshielded and placed at a point outside of but equivalent to the coldest part of the area to be protected in a position where it will not be damaged by farm equipment. It should also be shaded from morning sunlight.

With black currant blossom, which is particularly difficult to wet, sprinkling should commence when the falling air temperature reaches 0°C and should continue until the air temperature remains above 0°C and the ice starts melting on its own. With apples, it is possible under some conditions for the commencement of sprinkling to be delayed until the air temperature falls to −1°C, resulting in considerable savings in the use of water. Once started, however, the sprinklers should be kept in operation until the air temperature rises above 0°C.

An adjustable contact mercury thermometer wired through a relay can be arranged to ring a bell in the control room or operator's bedroom (the thermometer being set a little above freezing to allow time for the operator to reach the pump unit) or if required it can be used to set the sprinkler system in operation automatically. Two such adjustable thermometers can be used to provide a differential on/off temperature operation such as a −1°C start and 0°C stop. Warning equipment can also now be obtained to indicate, in the control room or bedroom, the actual temperature in the plantation.

The Meteorological Office provide a frost warning scheme for growers from the London Weather Centre. Warnings are issued by telephone at any time of the day or night when frosts are likely to occur, although with a normal radiation frost, they can usually be issued by about 5 p.m. Frost can be very local in occurrence and precise forecasts cannot be effected. Details of current charges can be obtained from the Meteorological Office.

WATER APPLICATION RATE

The particular crop, variety, stage of flower development, and other environment factors determine the amount of protection required and the rate of water application necessary to achieve this. Black currants are particularly vulnerable at the early grape stage and Bramley's Seedling and Cox's Orange Pippin apple at the early green cluster stage.

As the heat released depends on the amount of water that freezes on the plant, in theory the more severe the frost the greater the sprinkler application rate needs to be. In practice it is difficult to vary this and so for practical purposes a rate is chosen that will provide adequate protection against the normal radiation frosts occurring at this time. Present recommendations are based on a water application rate of at least $\frac{1}{8}$ in./hr which will give protection against 5°C of frost. With black currants a higher rate of $\frac{1}{7}$ in. per hour is desirable. If $\frac{1}{8}$ in./hr is to be applied a water flow of 47 gal/min is required for each acre covered and at $\frac{1}{7}$ in./hr 54 gal/min is required. Application rates that are too low can increase frost damage.

It is likely that crops under cloches or polythene tunnels can be successfully protected by frost sprinkling. In a preliminary trial using continuous sprinkler during frost over 'growers barn' type cloches the minimum temperature under the cloches was increased by 3°C and the period of frost reduced by about 6 hr/night.

Frost sprinkling over polythene tunnels is being further investigated. There is a suggestion with such crops under cover that the sensing thermometer can be located under the tunnels and used to automatically open a solenoid valve when the temperature drops. The resultant intermittent sprinkling would reduce the amount of water applied.

In assessing water requirements sufficient water should be available to provide for at least three and preferably more successive nights of frost each of 10 hr duration with provision for a total of 60 to 80 hr operation during the season. A saving of water can be made by taking advantage of any opportunity to recover the sprinkled water by running or pumping back into a reservoir.

OPERATION OF A SPRINKLER SYSTEM

Good equipment and a very high standard of maintenance is necessary to avoid breakdown. The equipment should be set up and tested well in advance of the first expected frost. It is advisable for an operator to be present throughout the period of sprinkling to ensure that the system is operating satisfactorily. Even with an automatic installation, pumps, motors, water pressures and sprinklers need to be checked. A powerful spot torch is useful to check each sprinkler at the start of sprinkling. Blocked sprinklers are best replaced by spares and this can be simplified by having standpipes incorporating automatic valves which seal when the sprinkler pipe is removed. If a pressure drop occurs during sprinkling and cannot be rectified it is advisable to shut off some lines rather than try to continue to cover the whole area at the lower pressure. Particular attention should be given to filters which should be cleaned regularly.

ANCILLARY USES OF SEMI-PERMANENT (SOLID SET) SPRINKLER SYSTEMS

In addition to frost protection and irrigation, a sprinkler installation of this kind can be used to apply nutrients by using an injector unit adjacent to the plantation or more simply by feeding the material into the suction side of the pump. Spray materials have also been applied in this way but normal spraying machines achieve better leaf cover and control of under leaf diseases.

It is claimed that fruit colour of apples can be improved later in the growing season by water sprinkling for about 10 min after relatively low night temperatures of 7° to 10°C on a morning or mornings when day temperatures are thought likely to rise to 18°C.

COSTS

In installing a solid set sprinkler system for frost protection, the entire area to be protected has to be covered by the equipment. Costs vary but can range between £300 and £650 per acre for the field equipment, pumps and mains. To this must be added the cost of any storage reservoir which may be required, which can vary between £600 to £1,500 per million gallons. Lined reservoirs can cost from three to five times as much depending on the type of lining.

Approved schemes are eligible for grant aid under the Ministry of Agriculture, Fisheries and Food Farm Water Supply Grant Scheme.

MULTI-PURPOSE IRRIGATION

The high capital costs of overhead irrigation have usually been minimized by having only enough equipment to irrigate a fraction of the whole area concerned, e.g., one fifth to one twentieth, and moving the equipment progressively from one side to another. There is, however, a trend towards the use of 'solid-set' installations, in which the whole area is permanently served by sprinklers and the equipment is not moved about. This is much more convenient and uses far less labour, but it is so much more costly than the movable system that probably it can be justified only if the equipment can be made to do several useful jobs besides irrigation. One obvious possibility is for protecting fruit against frost by sprinkling at night when orchard temperatures fall below the danger level. Another attraction of the solid-set system is in the application of fertilizers, fungicides, insecticides, leaf nutrients, growth substances, or perhaps even herbicides where these can be applied safely from overhead, though various technical problems need solving in this connection. A further possibility is that the equipment might be used to ameliorate the crop environment on hot, dry, windy days, when stomata may close to the detriment of growth, but this application is probably more important in hotter climates and little thought has been given to its managerial potential in Britain. An advantage of the multi-purpose system is that it can be fully automated, if desired, to switch on and off according to wind speed, temperature, gallonage to be applied, or virtually any other measurable attribute. The system is likely to be applicable only to high-value crops (fruit, flower crops, nursery beds, and so on) but might be of particular interest on light but well-drained soils if its use led to much heavier yields than such land would usually produce. The high cost of the installation could then be partly offset against the relatively low costs of what would otherwise be regarded as 'poor' land.

Appendix I

METEOROLOGICAL DATA

POTENTIAL TRANSPIRATION AVERAGES AND AVERAGE
MONTHLY DISTRIBUTION OF ANNUAL RAINFALL

For convenience, Scotland, Northern Ireland, England and Wales have been subdivided into 18 roughly homogeneous areas.

For each of the counties in each of these areas, the average potential transpiration is given for the mean county height above sea level and for the coastal area (where appropriate).

For each of the areas the 'rainfall percentages' are quoted; these percentages are given for each summer month (April–September) and for the winter (October–March) as a composite six-month percentage.

For example, Area 1, North Scotland

Month	Percentage	Average annual rainfall (say) 50 in.	Approximate averages
April	6.5		3.25
May	5.8	Multiply 50 by each	2.9
June	6.4	percentage in them	3.2
July	7.5	to obtain last column	3.75
August	8.4		4.2
September	9.6		4.8
October to March	55.8		27.9
		Total	50.0

Once the average annual rainfall is known, the monthly averages can be found by using these 'rainfall percentages' in this manner.

AVERAGE POTENTIAL TRANSPIRATION AND RAINFALL

Area		Counties
1.	N. Scotland	Caithness; Orkney; Ross and Cromarty; Sutherland; Zetland.
2.	W. Scotland	Argyll; Bute and Arran; Kintyre; Islay and Jura; Skye; Tiree; Coll and Mull; Uist; Barra and Harris.
3.	Central Scotland	Clackmannan; Dumbarton; Fife; Inverness; Kinross; Perth; Stirling.
4.	N.E. Scotland	Aberdeen; Angus; Banff; Kincardine; Moray; Nairn.
5.	S.W. Scotland	Ayr; Dumfries; Kirkcudbright; Lanark; Renfrew; Wigtown.
6.	S.E. Scotland	Berwick; E. Lothian; Midlothian; Peebles; Roxburgh; Selkirk; W. Lothian.
7.	N. Ireland	Antrim; Armagh; Down; Fermanagh; Londonderry; Tyrone.
8.	N.W. England	Cumberland; Isle of Man; Lancashire; Westmorland.
9.	N.E. England	Durham; Northumberland; East Riding; North Riding; West Riding.
10.	West Midlands	Cheshire; Gloucester; Hereford; Shropshire; Stafford; Warwick; Worcester.
11.	East Midlands	Derby; Leicester; Lincoln; Northampton; Nottingham; Rutland.
12.	E. England	Bedford; Cambridge; Essex; Hertford; Huntingdon; Norfolk; Suffolk.
13.	S.W. England	Cornwall; Devon; Scillies; Somerset.
14.	S. England	Berkshire; Buckingham; Dorset; Hampshire; Isle of Wight; Oxford; Wiltshire.
15.	S.E. England	Kent; Middlesex; Surrey; Sussex.
16.	N. Wales	Anglesey; Caernarvon; Denbigh; Flint; Merioneth.
17.	Central Wales	Brecon; Cardigan; Montgomery; Radnor.
18.	S. Wales	Carmarthen; Glamorgan; Pembroke; Monmouth.

POTENTIAL TRANSPIRATIONS

Height correction: add 0·2 in. to the summer total for each 100 ft below the county average height, spreading the correction in units of 0·05 over the monthly values; subtract the same amount, 0·2 in. for each 100 ft above the county average.

RAINFALL

Given the annual average rainfall for a given site, multiply this by the given monthly percentage to obtain the monthly averages for the site.

1. Scotland—North

	Average height	Apr.	May	June	July	Aug.	Sept.	Winter total
Caithness	360	2·00	2·85	3·15	2·85	2·25	1·50	1·70
Coast		2·15	2·90	3·25	3·00	2·35	1·65	2·80
Orkney	200	1·95	2·85	3·10	2·80	2·25	1·40	2·90
Coast		2·05	2·90	3·20	2·90	2·30	1·50	3·10
Ross and Cromarty E.	480	1·90	2·80	3·10	2·95	2·30	1·45	2·25
W.	490	1·90	2·80	3·10	2·90	2·25	1·45	1·60
E. Coast		2·30	3·15	3·50	3·30	2·45	1·70	3·25
W. Coast		2·15	3·20	3·35	3·00	2·50	1·75	3·15
Sutherland N.W.	480	1·85	2·80	3·10	2·90	2·25	1·40	1·50
Coast		2·15	3·10	3·30	3·00	2·40	1·70	2·90
S.E.	550	1·85	2·75	3·05	2·90	2·25	1·40	1·50
Coast		2·20	2·95	3·35	3·10	2·40	1·65	3·10
Zetland	210	1·90	2·75	3·05	2·85	2·25	1·40	2·95
Coast		2·00	2·85	3·15	2·90	2·30	1·50	3·10
Rainfall percentages		6·5	5·8	6·4	7·5	8·4	9·6	55·8

2. Scotland—West

County	Average height	Apr.	May	June	July	Aug.	Sept.	Winter total
Argyll	480	2·00	3·15	3·20	2·85	2·40	1·50	1·90
Coast		2·25	3·45	3·60	3·20	2·65	1·80	4·15
Bute and Arran	230	2·20	3·45	3·60	3·35	2·70	1·70	3·85
Coast		2·25	3·50	3·75	3·40	2·75	1·85	3·95
Kintyre, Islay and Jura	360	2·20	3·35	3·40	3·05	2·60	1·70	3·60
Coast		2·25	3·50	3·70	3·30	2·70	1·85	4·05
Skye	490	1·95	3·15	3·40	2·90	2·35	1·55	3·00
Coast		2·20	3·30	3·55	3·10	2·60	1·80	3·85
Tiree, Coll and Mull	340	2·20	3·30	3·35	3·00	2·55	1·70	4·15
Coast		2·35	3·45	3·60	3·10	2·65	1·85	4·70
Uist, Barra and Harris	260	2·15	3·30	3·60	3·05	2·55	1·75	3·40
Coast		2·25	3·40	3·65	3·10	2·65	1·85	3·70
Rainfall percentages		5·9	5·7	6·4	7·7	8·2	9·4	56·7

APPENDIX I

3. Scotland—Central

County	Average height	Apr.	May	June	July	Aug.	Sept.	Winter total
Clackmannan	270	2·10	3·00	3·45	3·40	2·55	1·60	2·35
Dumbarton	290	2·00	3·20	3·50	3·25	2·50	1·55	2·55
Fife E.	250	2·10	3·00	3·45	3·40	2·60	1·60	2·30
W.	370	2·05	2·95	3·40	3·30	2·55	1·55	2·20
Coast		2·20	3·10	3·55	3·50	2·70	1·70	2·95
Inverness E.	570	1·85	2·75	3·05	2·90	2·20	1·45	1·65
W.	490	1·95	3·15	3·40	2·90	2·35	1·55	3·00
Coast		2·20	3·30	3·55	3·10	2·60	1·80	3·85
Kinross	550	1·95	2·90	3·35	3·20	2·50	1·50	2·05
Perth	500	1·90	2·85	3·15	3·05	2·35	1·50	1·90
Stirling	330	1·95	3·20	3·45	3·25	2·50	1·55	2·15
Rainfall percentages		6·0	6·4	5·5	7·5	8·0	8·7	57·9

4. Scotland—N.E.

County	Average height	Apr.	May	June	July	Aug.	Sept.	Winter total
Aberdeen N.	260	2·00	2·90	3·20	3·05	2·35	1·60	1·95
S.E.	560	1·85	2·75	3·05	2·90	2·20	1·45	1·65
W.	810	1·75	2·65	2·95	2·80	2·10	1·35	1·40
Coast		2·20	3·00	3·30	3·20	2·50	1·80	3·65
Angus	420	2·00	2·90	3·20	3·10	2·40	1·60	1·90
Coast		2·20	3·05	3·45	3·40	2·65	1·70	3·05
Banff N.	420	2·10	3·00	3·45	3·20	2·30	1·50	2·50
S.	760	1·75	2·65	2·95	2·80	2·15	1·40	1·45
Coast		2·25	3·15	3·50	3·30	2·45	1·75	3·60
Kincardine	410	2·00	2·90	3·20	3·10	2·40	1·60	1·90
Coast		2·20	3·05	3·35	3·30	2·55	1·75	3·35
Moray	430	2·10	3·00	3·45	3·15	2·25	1·45	2·20
Coast		2·35	3·25	3·65	3·40	2·45	1·70	3·55
Nairn	470	2·15	3·05	3·50	3·20	2·25	1·45	2·40
Coast		2·35	3·25	3·65	3·40	2·45	1·70	3·60
Rainfall percentages		6·5	7·5	6·5	10·0	9·6	9·3	50·6

5. Scotland—S.W.

County	Average height	Apr.	May	June	July	Aug.	Sept.	Winter total
Ayr	500	2·00	3·25	3·50	3·20	2·30	1·50	2·50
Coast		2·25	3·50	3·75	3·40	2·75	1·85	3·90
Dumfries N.	730	1·80	3·05	3·25	2·85	2·25	1·50	2·00
S.	340	2·00	3·15	3·40	3·05	2·40	1·65	2·55
Coast		2·20	3·35	3·70	3·45	2·85	1·85	2·90
Kirkcudbright	450	2·00	3·15	3·40	3·10	2·40	1·65	2·45
Coast		2·25	3·35	3·75	3·45	2·90	1·90	3·35
Lanark	590	1·95	3·10	3·35	3·00	2·40	1·60	2·40
Renfrew	420	1·95	3·15	3·45	3·25	2·50	1·45	2·00
Wigtown	250	2·20	3·20	3·40	3·10	2·50	1·60	2·65
Coast		2·30	3·35	3·75	3·45	2·95	2·00	3·75
Rainfall percentages		5·6	6·2	6·0	7·5	8·5	9·4	56·8

6. Scotland—S.E.

County	Average height	Apr.	May	June	July	Aug.	Sept.	Winter total
Berwick	500	1·90	2·90	3·25	3·15	2·50	1·55	2·50
Coast		2·15	2·95	3·45	3·45	2·70	1·75	2·95
E. Lothian	350	2·05	3·00	3·45	3·40	2·60	1·60	2·10
Coast		2·20	3·00	3·50	3·45	2·70	1·70	2·95
Midlothian	580	1·95	3·05	3·40	3·15	2·45	1·60	2·40
Peebles	800	1·80	3·00	3·25	2·90	2·30	1·45	2·20
Roxburgh	610	1·85	2·85	3·20	3·10	2·50	1·50	2·40
Selkirk	760	1·80	2·80	3·20	3·10	2·45	1·45	2·25
West Lothian	420	2·00	3·05	3·40	3·20	2·55	1·60	2·40
Rainfall percentages		6·2	7·3	6·9	9·7	10·6	8·6	50·7

7. N. Ireland

County	Average height	Apr.	May	June	July	Aug.	Sept.	Winter total
Antrim	395	2·00	3·10	3·25	2·90	2·35	1·50	2·20
Coast		2·30	3·50	3·75	3·40	2·85	1·85	4·30
Armagh	325	2·05	3·15	3·25	2·90	2·35	1·50	2·65
Down	270	2·05	3·15	3·35	3·05	2·50	1·50	2·45
Coast		2·30	3·50	3·75	3·45	2·90	1·90	4·25
Fermanagh	370	2·05	3·10	3·25	2·90	2·35	1·45	2·30
Londonderry	340	2·05	3·15	3·30	2·90	2·40	1·55	2·55
Coast		3·25	3·50	3·70	3·35	2·80	1·85	4·55
Tyrone	410	2·00	3·15	3·25	2·90	2·35	1·45	2·30
Rainfall percentages		6·2	6·8	6·3	8·8	9·2	8·7	54·0

8. England—N.W.

County	Average height	Apr.	May	June	July	Aug.	Sept.	Winter total
Cumberland	415	1·95	3·10	3·40	3·30	2·60	1·60	2·20
Coast		2·15	3·35	3·65	3·45	2·85	1·80	2·90
Isle of Man	300	2·05	3·20	3·55	3·45	2·85	1·75	4·25
Coast		2·15	3·35	3·75	3·45	2·95	1·85	4·75
Lancashire	270	2·20	3·45	3·70	3·55	3·00	2·00	2·70
Coast		2·25	3·50	3·85	3·65	3·05	2·00	3·55
Furness		2·15	3·50	3·85	3·65	3·00	1·95	3·30
Westmorland	565	2·00	3·15	3·40	3·40	2·70	1·70	2·20
Rainfall percentages		5·7	6·2	6·2	9·0	10·2	9·7	53·0

APPENDIX I

9. England—N.E.

County	Average height	Apr.	May	June	July	Aug.	Sept.	Winter total
Durham	510	1·95	3·00	3·25	3·20	2·65	1·65	2·35
Coast		2·15	2·90	3·40	3·40	2·80	1·85	2·95
Northumberland	460	1·95	2·95	3·25	3·20	2·55	1·60	2·30
Coast		2·15	2·90	3·40	3·40	2·70	1·80	2·95
Yorkshire E.R.	150	2·10	3·15	3·50	3·40	2·75	1·80	2·60
Coast		2·10	3·05	3·45	3·45	2·85	1·95	2·90
N.R.	370	2·00	3·05	3·40	3·40	2·70	1·70	2·45
Coast		2·20	2·95	3·40	3·45	2·85	1·90	2·95
W.R.	385	2·05	3·10	3·50	3·50	2·80	1·75	2·40
Rainfall percentages		6·4	7·5	7·0	9·8	10·2	8·2	50·9

10. England—W. Midlands

County	Average height	Apr.	May	June	July	Aug.	Sept.	Winter total
Cheshire	220	2·15	3·25	3·65	3·60	3·00	1·90	2·75
Wirral		2·35	3·55	3·90	3·65	3·05	2·10	4·20
Gloucester	365	2·20	3·25	3·70	3·70	2·90	1·80	2·75
Hereford	405	2·20	3·05	3·70	3·65	2·75	1·75	2·70
Shropshire	430	2·10	3·10	3·60	3·55	2·85	1·85	2·55
Stafford	445	2·05	3·10	3·55	3·55	2·85	1·80	2·35
Warwick	330	2·10	3·10	3·70	3·70	2·95	1·90	2·50
Worcester	250	2·20	3·15	3·75	3·70	2·95	1·85	2·65
Rainfall percentages		6·9	8·2	6·3	9·0	8·8	8·0	52·8

11. England—East Midlands

County	Average height	Apr.	May	June	July	Aug.	Sept.	Winter total
Derby	475	2·05	3·10	3·40	3·30	2·60	1·60	2·20
Leicester	350	2·10	3·20	3·60	3·60	2·95	1·85	2·50
Lincoln	130	2·15	3·30	3·70	3·65	3·00	1·95	2·70
Coast		2·05	3·10	3·60	3·55	2·95	1·95	3·00
Northampton	350	2·15	3·15	3·70	3·70	3·00	1·90	2·50
Nottingham	185	2·25	3·30	3·65	3·60	2·90	1·85	2·65
Rutland	350	2·15	3·20	3·60	3·60	2·95	1·90	2·55
Rainfall percentages		7·2	7·8	6·8	10·0	9·5	8·2	50·5

12. England—East

County	Average height	Apr.	May	June	July	Aug.	Sept.	Winter total
Bedford	280	2·20	3·30	3·70	3·65	3·00	1·90	2·60
Cambridge	75	2·25	3·35	3·75	3·75	3·10	1·95	2·70
Essex	165	2·25	3·30	3·85	3·85	3·25	2·05	2·80
Coast		2·25	3·35	3·90	4·00	3·35	2·10	3·55
Hertford	340	2·15	3·25	3·70	3·70	3·00	1·90	2·70
Huntingdon	105	2·25	3·35	3·75	3·75	3·05	1·95	2·75
Norfolk	115	2·25	3·40	3·80	3·75	3·05	2·00	2·75
Coast		2·15	3·25	3·65	3·75	3·05	2·00	3·20
Suffolk	150	2·25	3·30	3·80	3·75	3·10	2·00	2·80
Coast		2·20	3·30	3·80	3·80	3·25	2·10	3·25
Rainfall percentages		7·5	7·5	6·9	10·2	8·8	9·0	50·1

13. England—S.W.

County	Average height	Apr.	May	June	July	Aug.	Sept.	Winter total
Cornwall	365	2·35	3·35	3·70	3·65	3·05	1·95	3·15
N. Coast		2·40	3·50	3·75	3·75	3·10	2·00	3·80
S. Coast		2·40	3·40	3·75	3·70	3·05	1·95	3·70
Devon	420	2·30	3·25	3·70	3·65	2·95	1·90	3·00
N. Coast		2·45	3·45	3·85	3·75	3·10	1·95	3·90
S. Coast		2·45	3·30	3·85	3·75	3·05	2·00	3·65
Scillies		2·55	3·50	3·75	3·80	3·20	2·15	5·85
Somerset	300	2·30	3·30	3·75	3·70	2·90	1·85	2·95
Coast		2·40	3·35	3·85	3·70	3·00	1·85	3·45
Channel Is.		2·50	3·65	3·85	4·40	3·65	2·40	6·30
Rainfall percentages		6·3	6·5	4·9	7·4	7·8	8·0	59·1

14. England—South

County	Average height	Apr.	May	June	July	Aug.	Sept.	Winter total
Berkshire	320	2·20	3·30	3·70	3·70	3·00	1·90	2·80
Buckingham	340	2·20	3·20	3·70	3·70	3·00	1·90	2·60
Dorset	295	2·25	3·25	3·80	3·75	2·95	1·80	2·90
Coast		2·40	3·30	3·85	3·85	3·25	2·10	3·60
Hampshire	260	2·20	3·20	3·75	3·75	3·05	1·85	2·80
Coast		2·30	3·30	3·90	4·00	3·35	2·00	3·60
Isle of Wight	185	2·25	3·35	3·80	3·80	3·20	2·05	3·45
Coast		2·45	3·55	3·95	4·00	3·35	2·10	3·75
Oxford	365	2·15	3·20	3·70	3·70	2·95	1·90	2·75
Wiltshire	400	2·10	3·25	3·75	3·75	2·95	1·80	2·70
Rainfall percentages		7·0	7·3	5·7	8·2	8·0	8·2	55·6

APPENDIX I

15. England—S.E.

County	Average height	Apr.	May	June	July	Aug.	Sept.	Winter total
Kent	240	2·25	3·25	3·95	3·90	3·15	2·00	2·75
N. Coast		2·25	3·35	4·00	4·10	3·35	2·10	3·55
S. Coast		2·30	3·30	3·90	4·10	3·45	2·10	3·70
Middlesex	135	2·25	3·25	3·85	3·80	3·10	2·00	3·10
Surrey	280	2·20	3·20	3·80	3·75	3·10	1·90	2·90
Sussex	210	2·20	3·20	3·75	3·75	3·10	1·90	2·90
Coast		2·30	3·30	3·90	4·00	3·45	2·10	3·55
Rainfall percentages		7·1	6·5	5·6	8·1	8·8	7·5	56·4

16. Wales—North

County	Average height	Apr.	May	June	July	Aug.	Sept.	Winter total
Anglesey	165	2·15	3·25	3·50	3·45	2·85	1·95	3·45
Coast		2·15	3·30	3·60	3·45	2·85	1·95	3·70
Caernarvon	400	2·15	3·15	3·40	3·35	2·80	1·85	3·10
Coast		2·20	3·25	3·60	3·45	2·85	1·95	3·85
Denbigh	545	2·10	3·10	3·40	3·35	2·70	1·85	2·60
Coast		2·25	3·40	3·75	3·55	2·95	2·05	4·05
Flint	350	2·05	3·10	3·55	3·50	2·95	1·85	2·70
Coast		2·25	3·40	3·75	3·55	2·95	2·05	4·05
Merioneth	590	2·15	3·15	3·45	3·35	2·80	1·90	2·60
Coast		2·20	3·30	3·60	3·45	2·85	1·95	4·15
Rainfall percentages		5·7	5·7	6·4	7·5	8·7	9·3	56·7

17. Wales—Central

County	Average height	Apr.	May	June	July	Aug.	Sept.	Winter total
Brecon	690	2·10	3·10	3·30	3·25	2·75	1·80	2·55
Cardigan	570	2·15	3·15	3·45	3·40	2·80	1·90	2·85
Coast		2·20	3·30	3·65	3·45	2·85	2·00	4·30
Montgomery	640	2·15	3·10	3·40	3·35	2·75	1·85	2·55
Radnor	770	2·10	3·10	3·35	3·30	2·70	1·80	2·45
Rainfall percentages		6·5	6·4	5·9	7·7	8·4	8·4	56·8

18. Wales—South

County	Average height	Apr.	May	June	July	Aug.	Sept.	Winter total
Carmarthen	445	2·15	3·15	3·60	3·60	2·90	1·90	2·95
Coast		2·30	3·40	3·85	3·65	3·05	1·95	3·80
Glamorgan	415	2·15	3·15	3·70	3·60	3·05	1·90	2·70
Coast		2·30	3·40	3·85	3·60	3·10	1·90	3·60
Pembroke	310	2·15	3·30	3·60	3·50	3·00	1·95	3·25
Coast		2·20	3·40	3·75	3·60	3·05	2·15	3·85
Monmouth	410	2·15	3·15	3·65	3·60	2·95	1·90	2·70
Rainfall percentages		5·8	5·8	5·4	8·2	9·0	8·6	57·2

Appendix II

IRRIGATION—ESTIMATION OF FREQUENCY, MAXIMUM AND AVERAGE NEED

THE diagrams which follow enable information regarding frequency of need, irrigation maximum and average need to be estimated from details of average rainfall and average potential transpiration for the four plans detailed on page 12 of this Bulletin.

A numerical link can be used between the irrigation need for the 5th driest year in 20 (x) and the total (adjusted) need in 20 years (which under-irrigates the four driest years), namely

Plan	Total adjusted need
Plan 1	$15 \cdot 6x - 12 \cdot 2$
Plan 2	$15 \cdot 6x - 12 \cdot 8$
Plan 3	$15 \cdot 6x - 16$
Plan 4	$15 \cdot 6x - 34$

If an estimate of the total need in 20 years is required, this can be found by multiplying the total adjusted need by the following factors:

Total adjusted need	Plan 1	Plan 2	Plan 3	Plan 4
20	1·20	1·20	1·17	1·40
30	1·15	1·15	1·14	1·30
40	1·10	1·10	1·12	1·20
50	1·06	1·08	1·09	1·15
60 in. or more	1·06	1·06	1·07	1·10

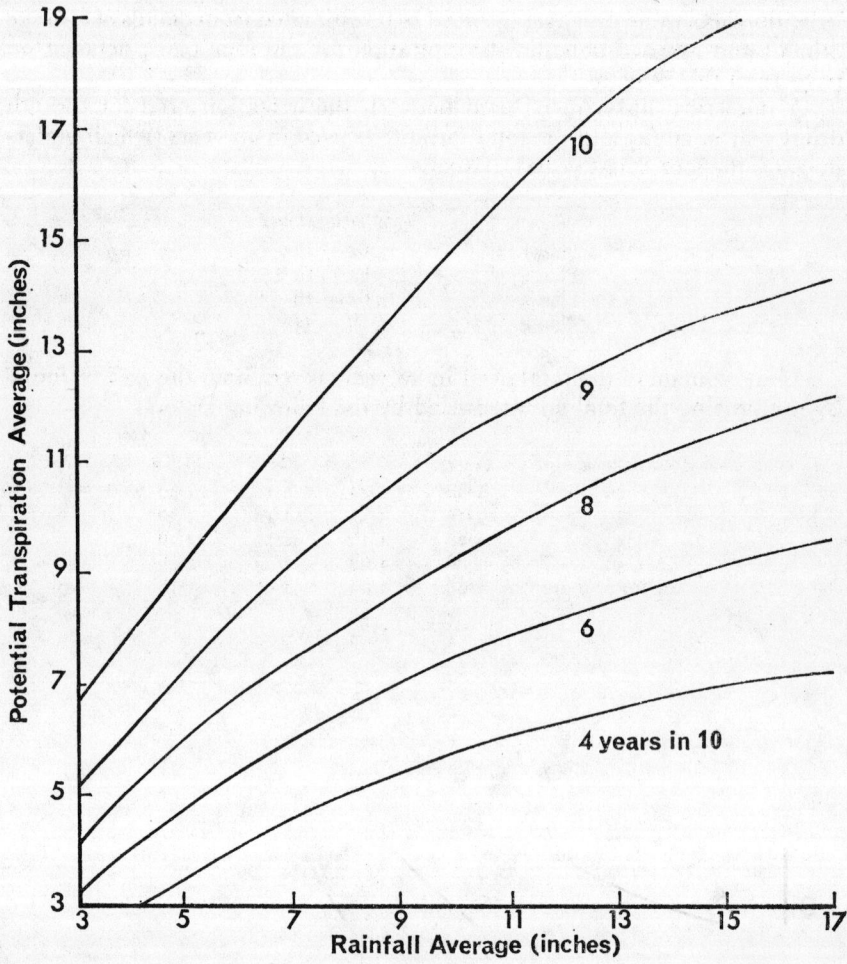

Fig. 24. Frequency of irrigation need, Plan 1. Soil restored to capacity whenever soil moisture deficit becomes 1 in.

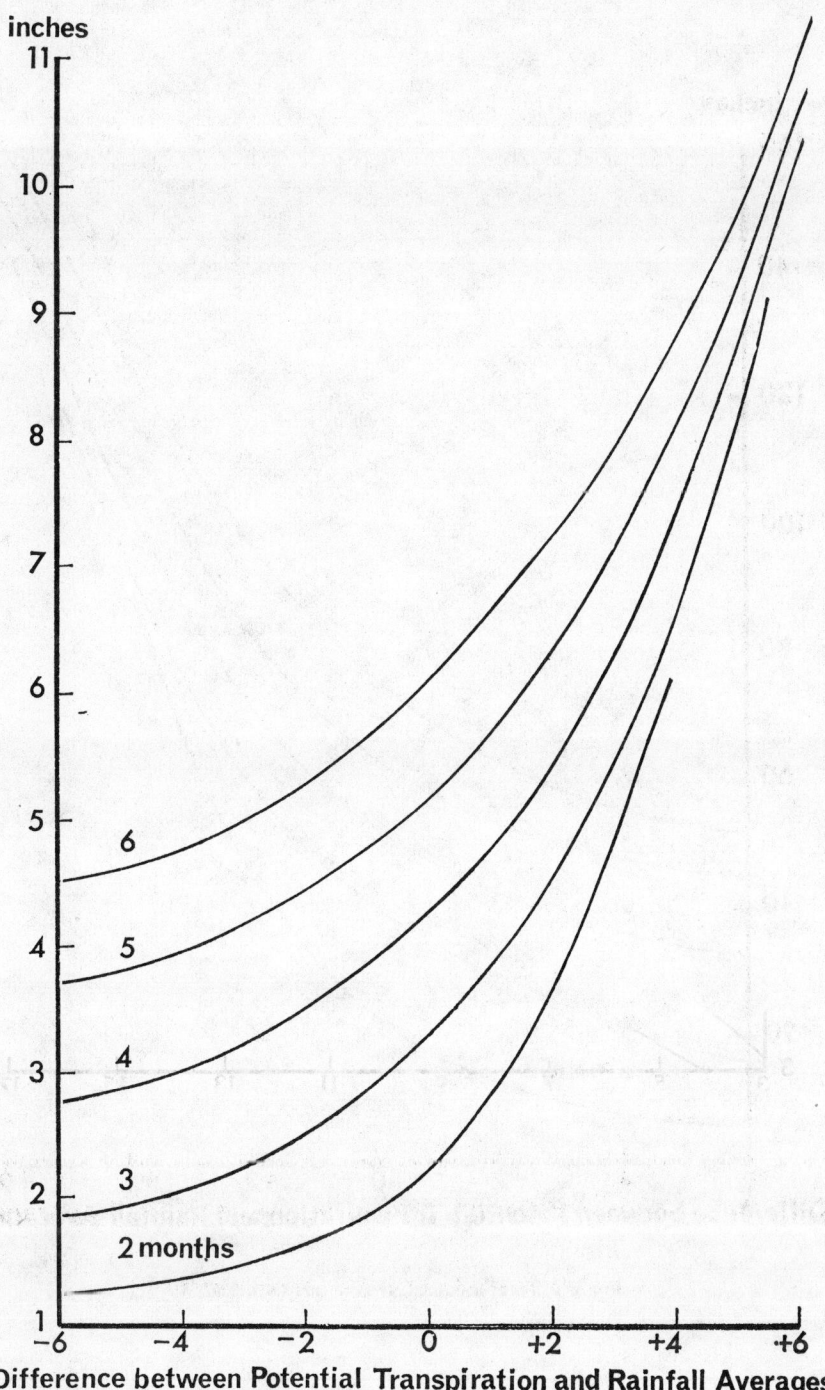

Fig. 25. Irrigation need for fifth driest year in 20. Plan 1

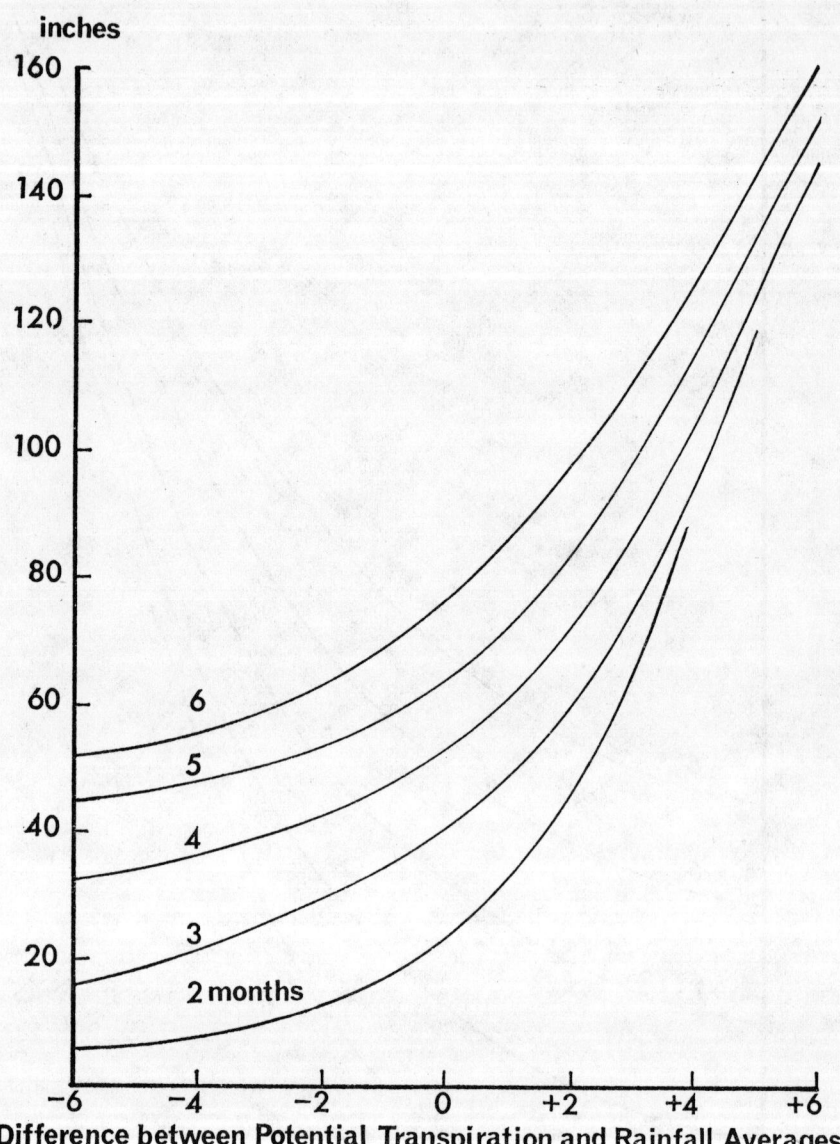

Fig. 26. Total adjusted need in 20 years. Plan 1

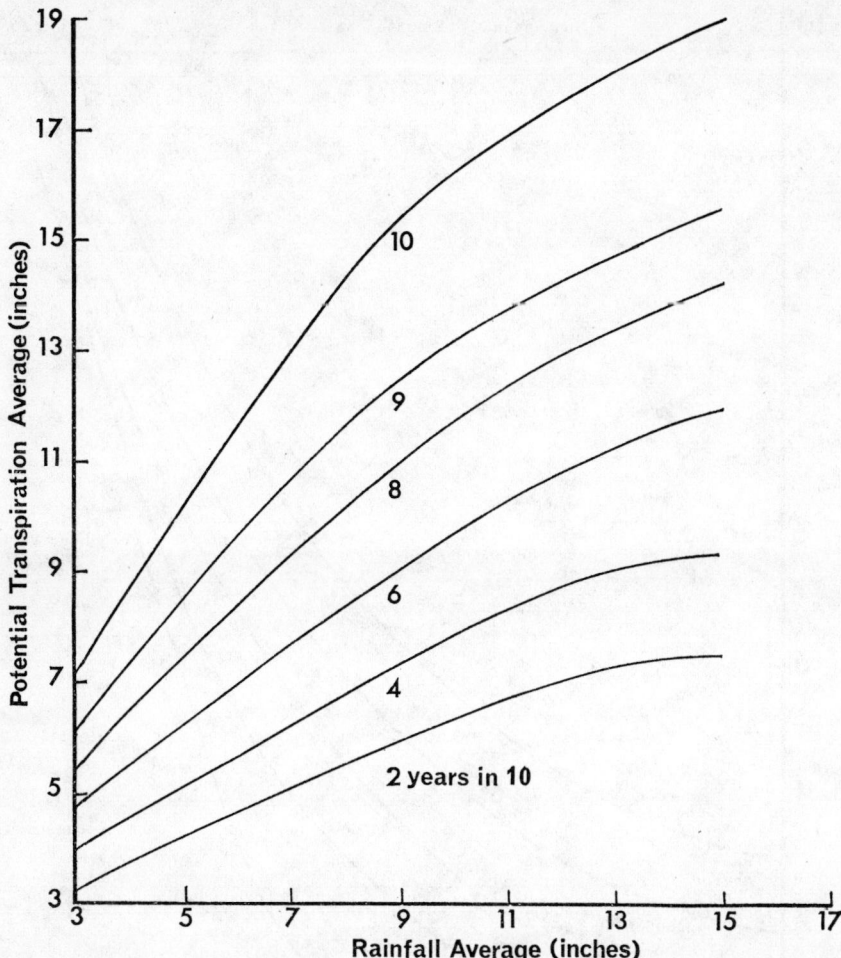

Fig. 27. Frequency of irrigation need, Plan 2. Soil restored to capacity whenever soil moisture deficit becomes 2 in.

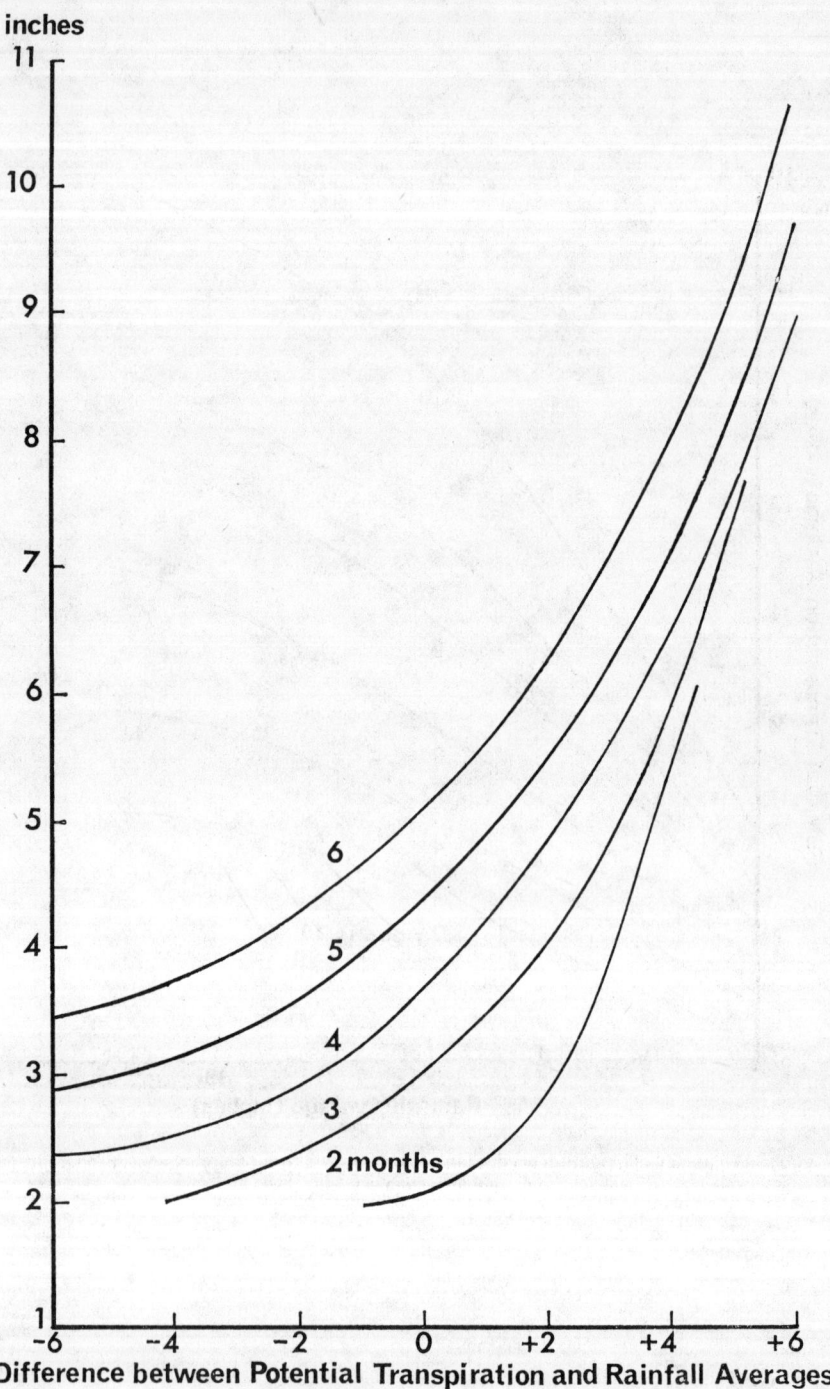

Fig. 28. Irrigation need for fifth driest year in 20. Plan 2

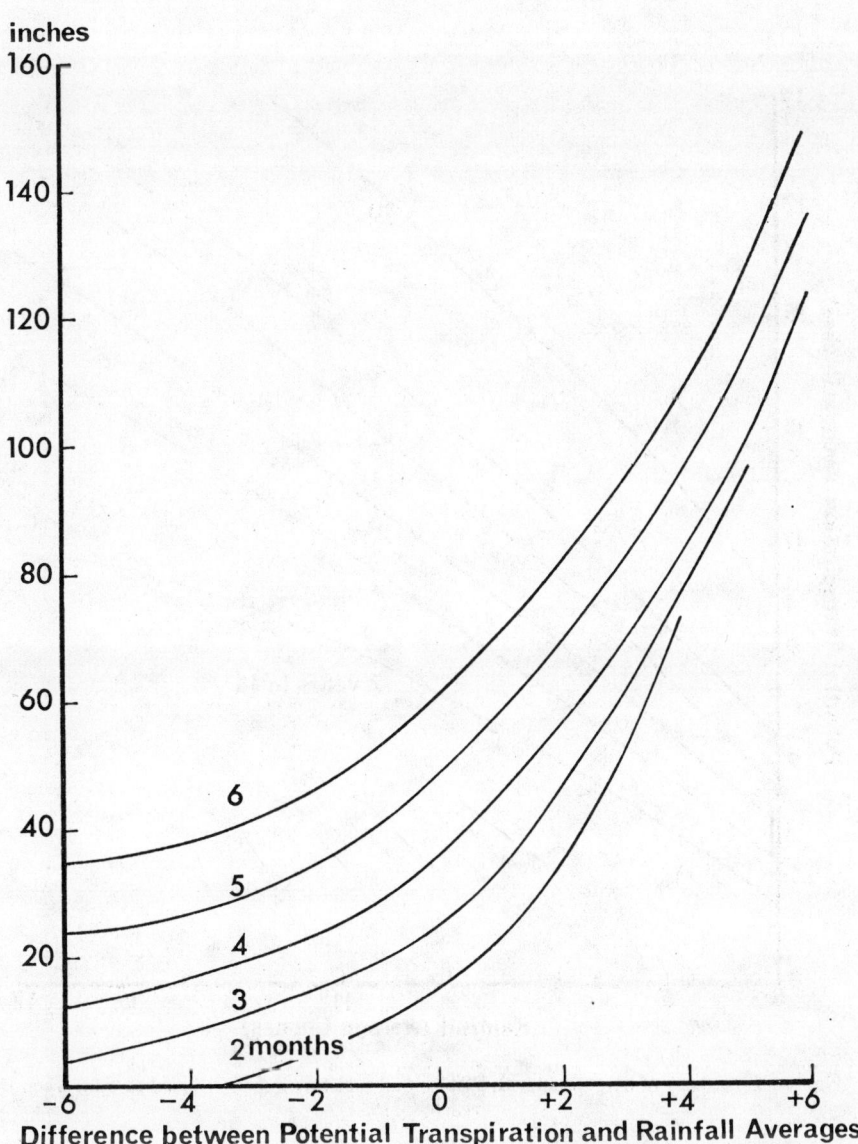

Fig. 29. Total adjusted need in 20 years. Plan 2

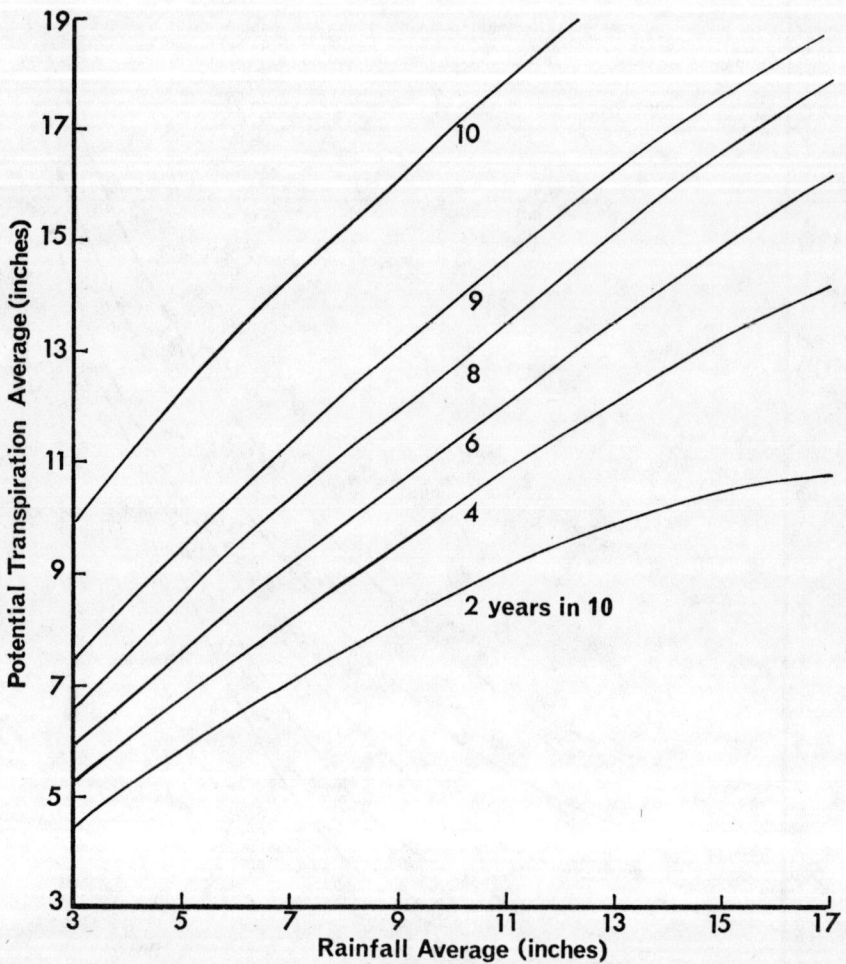

Fig. 30. Frequency of irrigation need, Plan 3. Soil moisture deficit reduced to 1 in. when it reaches 3 in.

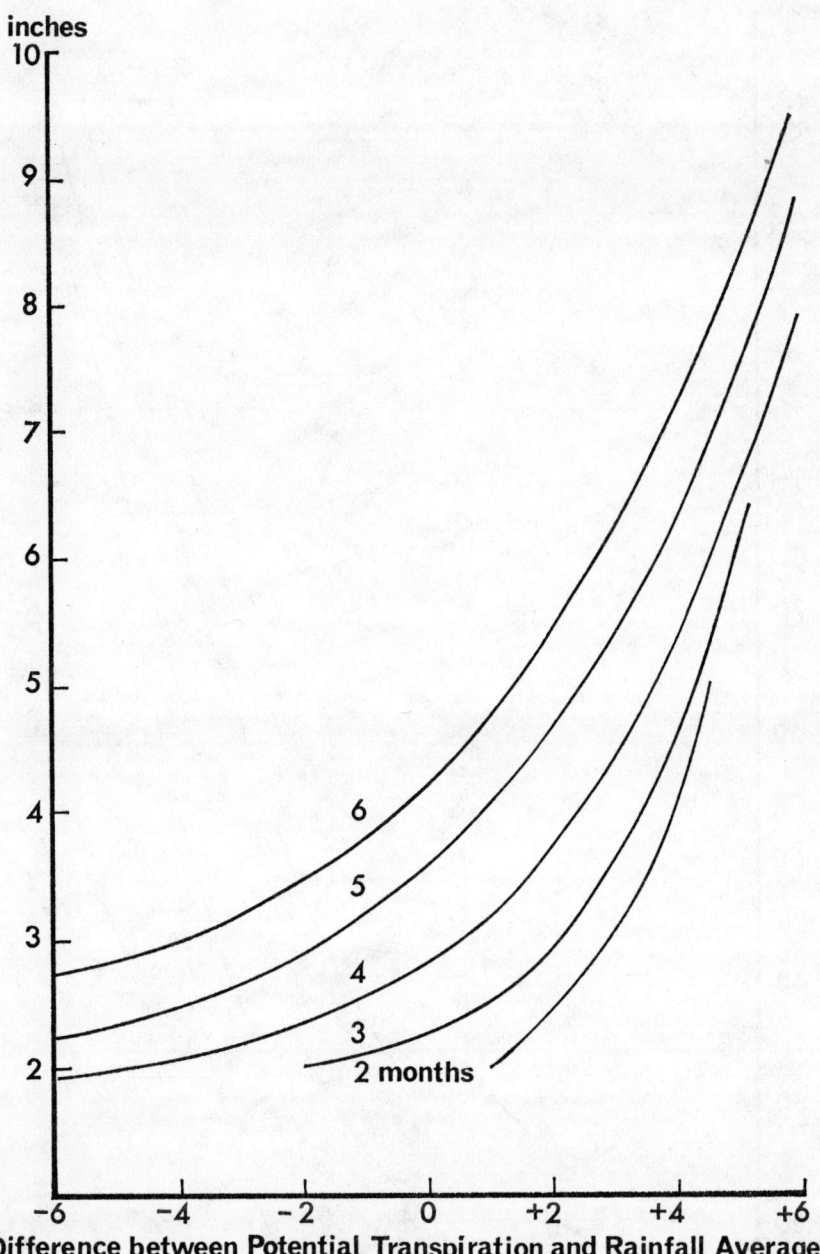

Fig. 31. Irrigation need for fifth driest year in 20. Plan 3

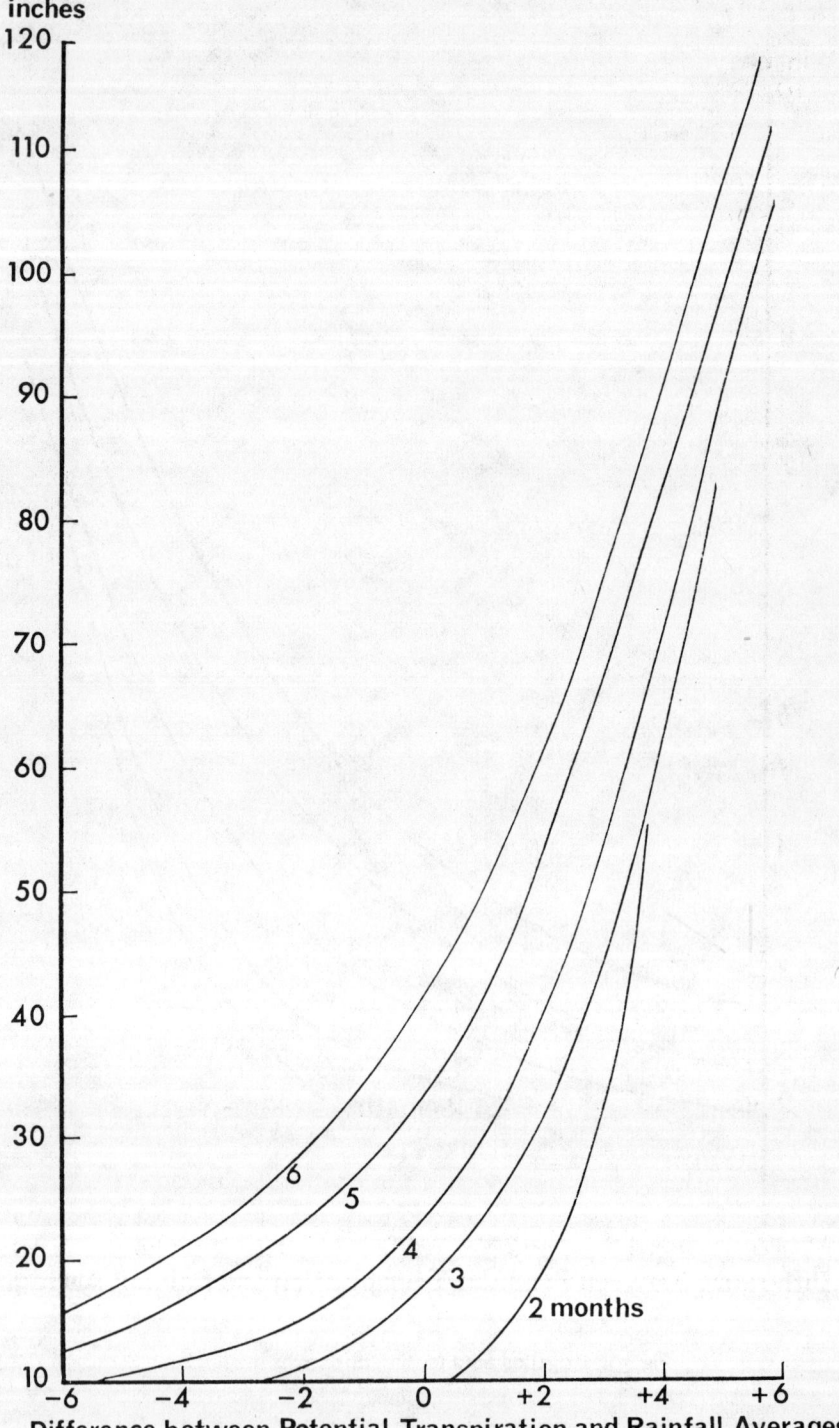

Fig. 32. Total adjusted need in 20 years. Plan 3

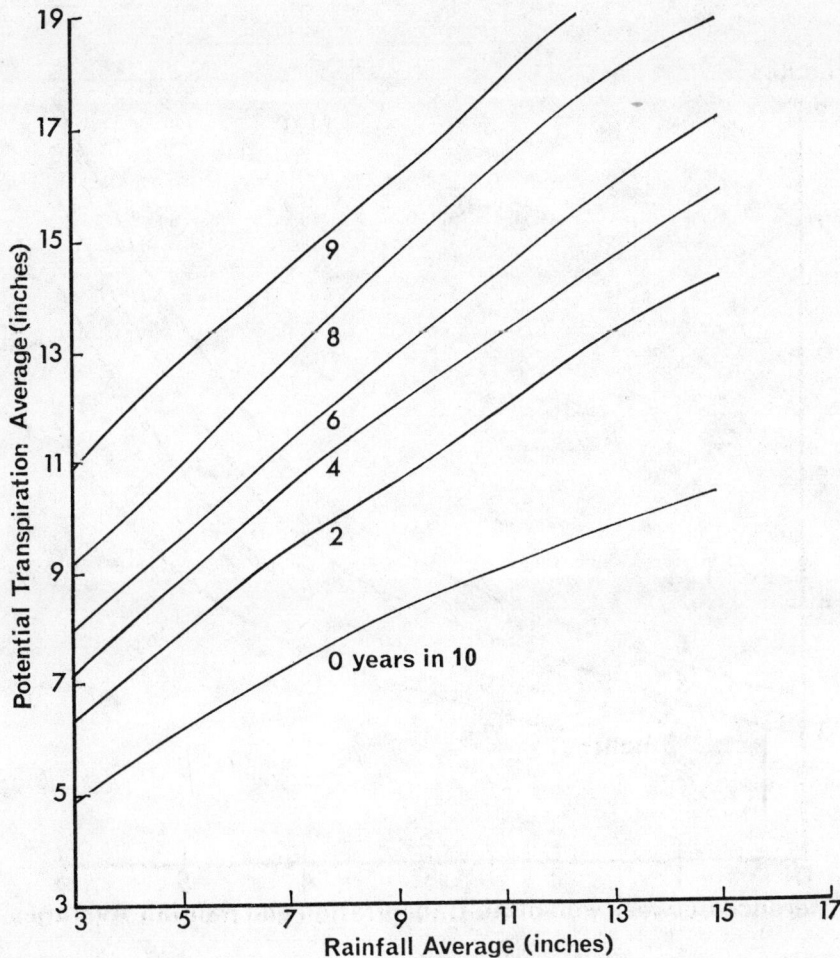

Fig. 33. Frequency of irrigation need, Plan 4. Soil moisture deficit reduced to 2 in. when it reaches 5 in.

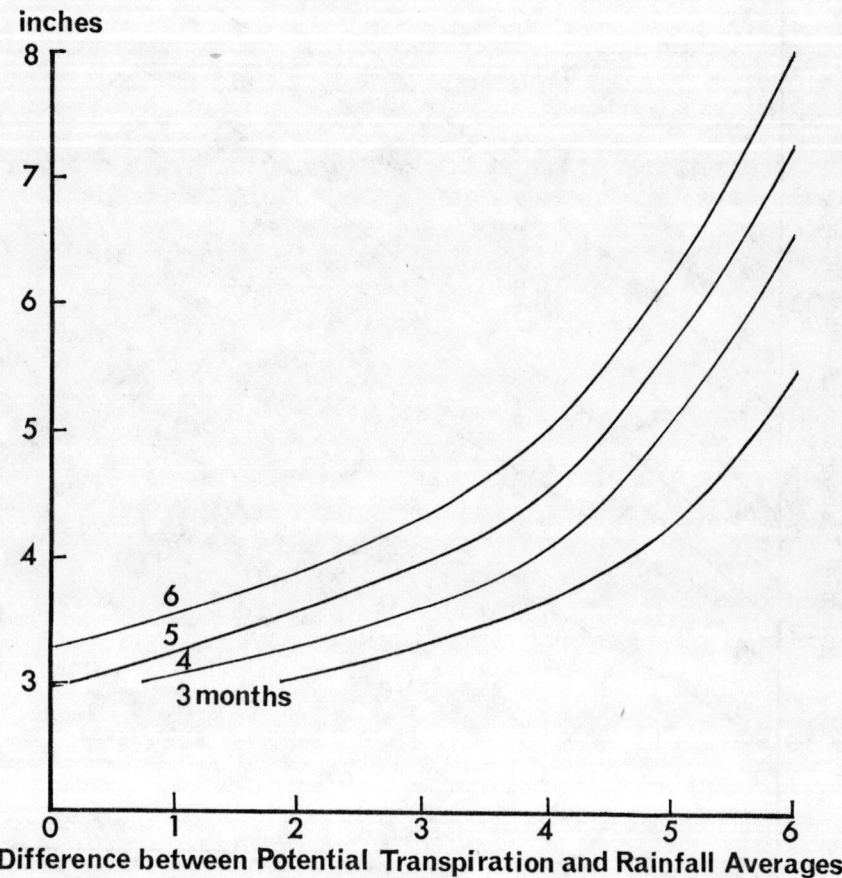

Fig. 34. Irrigation need for fifth driest year in 20. Plan 4

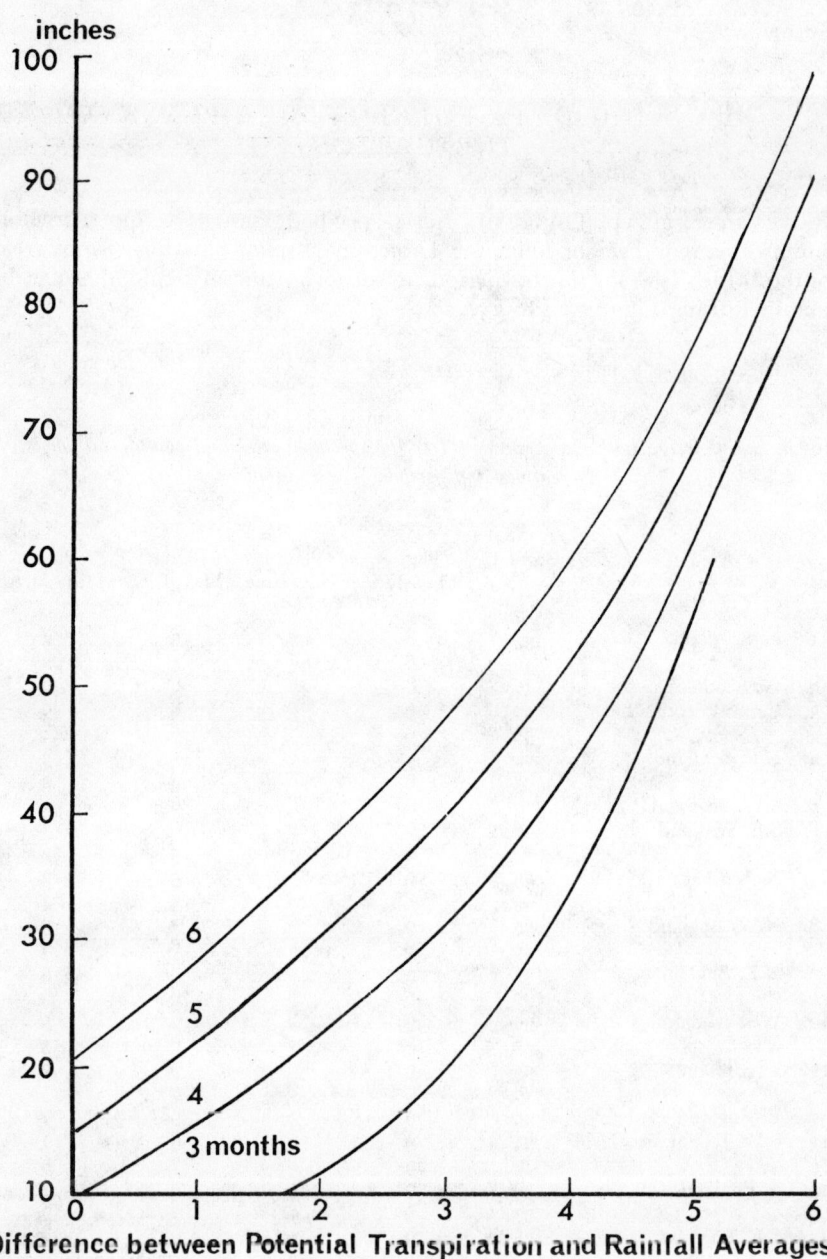

Fig. 35. Total adjusted need in 20 years. Plan 4

Appendix III

PLANNING DATA FOR SPECIMEN PLANS OF IRRIGATION

PAGES 12–14 give planning data for 4 specified Plans covering the whole summer six months. The following Tables give similar data for various areas of the British Isles for shorter periods to show the kind of results that can be obtained using Figures 24 to 35.

Frequency of irrigation need (years in 10) for specimen sites and sample soil moisture regimes and periods of irrigation

	Area	Annual Rainfall Average (in.)	Plan 1 May–June	Plan 2 May–June	Plan 3 May–July	Plan 4 May–August
1.	N. Scotland	35	8	7	5	2
		40	8	6	4	2
2.	W. Scotland	40	8	7	6	2
3.	Central Scotland	30	9	8	8	6
		35	9	7	6	4
4.	N.E. Scotland	30	9	7	5	3
		35	8	6	4	2
5.	S.W. Scotland	40	8	6	6	2
6.	S.E. Scotland	30	9	8	6	3
		35	8	7	5	2
7.	N. Ireland	35	9	7	5	2
		40	8	6	4	2
8.	N.W. England	35	9	8	7	3
		40	8	7	6	2
9.	N.E. England	25	9	8	8	5
		30	9	8	6	3
10.	West Midland	30	9	8	7	5
		35	8	7	6	3
11.	E. Midlands	25	9	9	9	7
		30	9	8	7	5
12.	E. England	25	9	9	9	7
13.	S.W. England	35	9	9	8	7
		40	9	8	7	5
14.	S. England	30	9	9	9	7
		35	9	8	7	6
15.	S.E. England	25	10	9	9	8
		30	9	9	9	6
16.	N. Wales	30	9	8	8	6
		35	9	7	7	4
17.	Central Wales	35	9	8	7	4
		40	8	7	6	2
18.	S. Wales	35	9	9	8	5
		40	9	8	7	3

APPENDIX III

Amount of irrigation need for specimen site and sample soil moisture regimes and periods of irrigation

	Area	Annual Rainfall Average (in.)	Plan 1 May–June (a)	(b)	Plan 2 May–June (a)	(b)	Plan 3 May–July (a)	(b)	Plan 4 May–August (a)	(b)
1.	N. Scotland	35	3·4	40	2·7	30	2·0	33	3·2	18
		40	2·9	32	2·4	23	2·6	25	2·8	10
2.	W. Scotland	40	3·5	42	2·8	32	2·8	29	3·1	16
3.	Central Scotland	30	4·6	60	3·8	47	4·5	56	4·2	42
		35	4·0	52	3·2	38	3·6	41	3·5	27
4.	N.E. Scotland	30	3·6	45	2·9	33	3·0	33	3·3	19
		35	3·0	34	2·5	25	2·5	23	2·8	10
5.	S.W. Scotland	40	3·4	40	2·7	30	3·0	33	3·2	17
6.	S.E. Scotland	30	3·9	50	3·1	36	3·4	37	3·3	22
		35	3·2	38	2·6	27	2·6	26	2·9	11
7.	N. Ireland	35	3·5	42	2·8	32	2·7	28	3·1	15
		40	3·0	34	2·5	25	2·4	20	2·6	8
8.	N.W. England	35	4·3	55	3·4	43	3·5	39	3·4	23
		40	3·6	45	2·9	33	2·7	29	3·0	16
9.	N.E. England	25	4·6	60	3·8	47	4·2	52	4·1	42
		30	3·8	48	3·0	35	3·4	37	3·4	24
10.	W. Midland	30	4·1	54	3·3	40	3·9	46	3·8	37
		35	3·3	39	2·7	29	3·0	33	3·3	21
11.	E. Midlands	25	5·0	68	4·3	53	5·1	66	5·2	64
		30	4·3	55	3·4	43	3·8	44	3·8	37
12.	E. England	25	5·5	75	4·5	59	5·3	70	5·9	75
13.	S.W. England	35	4·8	63	4·0	50	4·9	63	4·5	48
		40	4·2	54	3·4	42	3·8	45	3·7	31
14.	S. England	30	4·8	63	4·0	50	5·0	64	5·0	58
		35	4·1	53	3·3	40	3·9	47	3·8	37
15.	S.E. England	25	6·0	80	5·4	70	7·4	90	7·5	90
		30	5·3	71	4·7	60	6·3	80	5·7	70
16.	N. Wales	30	4·5	58	3·7	46	4·7	60	4·2	45
		35	3·8	48	3·0	34	3·6	41	3·5	28
17.	Central Wales	35	4·0	52	3·2	38	3·6	41	3·6	29
		40	3·3	39	2·7	29	2·9	31	3·2	18
18.	S. Wales	35	4·7	61	3·9	48	4·4	54	3·8	37
		40	4·1	54	3·3	39	3·4	38	3·4	23

(a) = Need (in.) in 5th driest year in 20.
(b) = Total need (in.) in 20 years, adjusted for driest years to maximum of 5th driest.

The following pages explain how knowledge of average rainfall and transpiration and their variations from year to year can be used to obtain planning data for any irrigation regime other than those specified in the four quoted Plans.

Appendix IV

ADDITIONAL IRRIGATION PLANS

ESTIMATION of irrigation needs for plans other than the four already specified needs a lengthier procedure. Suppose that it was required to find the frequency for a plan which demanded:

Deficit reduced to 2 in. when it reaches 4 in. throughout June, July and August, for a 25 in. annual average rainfall site in Suffolk 125 ft above mean sea level.

First it is necessary to estimate the soil moisture deficit at the end of June. From the Appendix, Area 12 (England, East), the potential transpiration at 150 ft a.m.s.l. can be found. A height correction must be made, adding 0·05 in. for the 25 ft below the county average height; this can conveniently be added to the July average. The April–May total is therefore $2·25+3·30 = 5·55$ in. Remembering that a crop such as sugar beet or potatoes will not be covering much of the ground in April, the average water used during these two months could be taken as 5 in.

Next, the average rainfall for these months is required; the rainfall percentages (of annual) are given as 7·5 for April and 7·5 for May, therefore 15 per cent of the 25 in. annual average total is 3·75 in.

The variations about such an average that might be expected in 10 sample years can be obtained from the first Table in this Appendix, *Estimated rainfall totals*. The sample percentages of the average 3·75 in. are given in section (b) along the 2 months line; namely

$39\frac{1}{2}$, 58, $70\frac{1}{2}$, $81\frac{1}{2}$, $92\frac{1}{2}$, 103, $113\frac{1}{2}$, $125\frac{1}{2}$, 142 and 174 per cent.

This gives sample totals of:

1·5, 2·2, 2·65, 3·05, 3·45, 3·85, 4·25, 4·7, 5·3 and 6·5 in.

Subtracting from the estimated 5 in. transpiration loss, the following estimates of end-of-May deficits are obtained:

3·5, 2·8, 2·35, 1·95, 1·55, 1·15, 0·75, 0·3, zero and zero.

Considering now the June to August period, the average potential transpirations from the Appendix are:

$3·80+3·80+3·10 = 10·7$ in.

The rainfall percentages of annual total are:

$6·9+10·2+8·8 = 25·9$ per cent

For 25 in. annual averages this implies a 3 month average rainfall of:

$25·9 \times 25 \div 100 = 6·5$ in.

The sample percentages of Table 2 (b) for 3 months are:

49, 66, $76\frac{1}{2}$, 85, 94, 102, $111\frac{1}{2}$, 121, $135\frac{1}{2}$, $159\frac{1}{2}$ per cent

giving sample rainfalls:

3·2, 4·3, 5·0, 5·5, 6·1, 6·6, 7·2, 7·9, 8·9, and 10·4 in.

APPENDIX IV

therefore by subtraction the sample (additional to April–May) potential moisture deficits are:

7·5, 6·4, 5·7, 5·2, 4·6, 4·1, 3·5, 2·8, 1·8 and 0·3 in.

In theory any one of these 10 samples could occur after any one of the previous 10 samples for April–May, giving 10 × 10 = 100 samples. In practice it is clear than six of the years, with potential deficits, June–August, ranging from 7·5 to 4·1 already indicate an irrigation need within the plan. Considering the remaining four years we have 40 sample end of August potential deficits.

7·0, 6·3, 5·85, 5·54, 5·05, 4·65, 4·25, 3·8, 3·5 and 3·5 in.
6·3, 5·6, 5·15, 4·75, 4·35, 3·95, 3·55, 3·1, 2·8 and 2·8 in.
5·3, 4·6, 4·15, 3·75, 3·35, 2·95, 2·55, 2·1, 1·8 and 1·8 in.
3·8 and nine others below 4·0 in.

giving a total of 15 out of 40, or 1·5 out of 4.

Estimated rainfall totals
(expressed as a percentage of average)

Length of period during April–September	Driest year	Next driest	3	4	5	6	7	8	Next wettest	Wettest year
(a) England–Northern Areas (north of a line Birmingham to Yarmouth)										
1 month	24½	47	62	75	87	99	113½	131½	155½	205
2 months	44	61	72	82	92	101	112½	123	139½	172
3 months	53	67	77	86	94	102	110	119	133	159
4 months	58½	71	79½	87½	94½	102	108	116½	130	152½
5 months	61	73	81	89	96	103	108½	116	126½	146
6 months	62½	74½	83½	91	97	101½	107½	115	123½	144
(b) England–Southern Areas (south of a line Birmingham to Yarmouth)										
1 month	20½	42½	59	73½	87	101	116½	134¼	159	206½
2 months	39½	58	70½	81½	92½	103	113½	125½	142	174
3 months	49	66	76½	85	94	102	110½	121	135½	159½
4 months	54	69½	79	87	94½	102	110	120	131	153
5 months	57½	71½	82	89	96	102½	109	117½	127½	147½
6 months	58½	72½	82½	90	96	103½	109½	116	127½	144
(c) Eastern Scotland										
1 month	31	50	62	74	86	98	111	129	152	207
2 months	45	61	74	84	93	101	110	122	138	172
3 months	54	68	79	87	95	101	109	118	132	157
4 months	62	74	81	89	95	101	108	116	127	147
5 months	65	77	83	90	95	101	107	114	124	144
6 months	67	78	84	90	95	101	107	114	122	142
(d) Western Scotland and Northern Ireland										
1 month	37	59	70	79	91	102	114	127	143	178
2 months	53	69	78	85	94	101	109	121	133	157
3 months	63	76	82	89	95	101	107	116	126	145
4 months	70	81	86	91	95	101	106	113	120	137
5 months	74	82	87	92	96	101	106	111	117	134
6 months	76	83	88	92	96	101	105	110	116	133

The estimated need for carrying out this plan is therefore $6+1 \cdot 5 = 7 \cdot 5$ years in 10. This is probably an underestimate owing to the fact that it cannot include those years in which there occurred a 4 in. potential deficit in the middle of the 3 month period which corrected by subsequent rainfall, in other words it omits temporary dry spells which may be significant. An answer of 8 years in 10 is therefore probably reasonable.

Using the same example, the next Table has been drawn up combining the 10 sample April–May deficits with the 10 sample June–August deficits.

Sample potential deficits April–August in Suffolk
(25 in. annual average rainfall, 125 ft a.m.s.l.)

June to August	\	\	\	April and May	\	\	\	\	\	Total adjusted need
	3·5	2·8	2·35	1·95	1·55	1·15	0·75	0·3	0	0	
7·5	11·0	10·3	9·85	9·45	9·05	8·65	8·25	7·8	7·5	7·5	55·00
6·4	9·9	9·2	8·75	8·35	7·95	7·55	6·95	6·7	6·4	6·4	51·45
5·7	9·2	8·5	8·05	7·65	7·25	6·85	6·45	6·0	5·7	5·7	47·95
5·2	8·7	8·0	7·55	7·15	6·75	6·35	5·95	5·5	5·2	5·2	44·60
4·6	8·1	7·4	6·95	6·55	6·15	5·75	5·35	4·9	4·6	4·6	39·75
4·1	7·6	6·9	6·45	6·05	5·65	5·25	4·85	4·4	4·1	4·1	35·25
3·5	7·0	6·3	5·85	5·54	5·05	4·65	4·25	3·8	3·5	3·5	24·55
2·8	6·3	5·6	5·15	4·75	4·35	3·95	3·55	3·1	2·8	2·8	16·15
1·8	5·3	4·6	4·15	3·75	3·35	2·95	2·55	2·1	1·8	1·8	8·05
0·3	3·8	3·1	2·65	2·25	1·85	1·45	1·05	0·6	0·3	0·3	0·00
											322·75

Need in fifth driest year in 20 = 5·5 in. Total adjusted need in 20 years = 65 in.
Note: The total adjusted need in the last column is obtained as follows:
  line 1. 10 years each needing the maximum of 7·5 = 2 − 5·5 in.
  line 2. 6 years with the maximum of 5·5 plus 4 years with 4·95, 4·7, 4·4 and 4·4 respectively—and so on.
  line 10. No year with 4 in. deficit, therefore the need is zero.

The 25 combined deficits to the top left-hand corner of this table, are all greater than the 5th driest year in 20 (which occurs with a 7·5 in. potential deficit). The deficits in the bottom right-hand corner, which also number 25, are those below 4·0 in., and so, according to the plan, will not require irrigation.

The right-hand column of this Table gives the estimated totals of adjusted irrigation, counting all the driest 25 per cent as the same requirement. This indicates that the need for the fifth driest year in 20 as

$$7 \cdot 5 - 2 \cdot 0 = 5 \cdot 5$$

as the deficit is 7·5 and the plan demands restoration to 2 in. The total adjusted need for 20 years is

$$\frac{322 \cdot 75}{5} = 65 \text{ in.}$$

The full need for 20 years would work out at 70 in., with a maximum requirement of 9 in. in the driest year of the century. This total need is approximately 8 per cent greater than the adjusted need.

Appendix V

IRRIGATION COSTS (1973)

A farmer contemplating the installation of irrigation equipment should study both the costs that are to be incurred and the benefits that are likely to be obtained under his own particular circumstances; the following details can only be given as a guide and no allowance has been made for grant:

Capital Costs

These can vary between very wide limits and depend on a number of factors such as:

(a) the type of system chosen (application equipment, permanent or temporary mains, type of pumping equipment);
(b) the size of the scheme;
(c) the accessibility of suitable water;
(d) the source, works and water storage required;
(e) the shape and topography of the land to be irrigated.

For agricultural systems a broad range of costs could be taken from £50 per acre (tractor pump scheme with water adjacent to irrigation area) to £150 per acre (permanent pump unit, underground main, with reasonable distances from water to irrigation area). Two typical layouts giving the approximate component (1973) costs are as follows:

(a) Tractor pump scheme with portable mains covering 35 acres in 10 days

	£
Tractor pump	320
600 ft of 4 in. portable main	450
1,230 ft of 3 in. sprinkler line	820
	1,590 or £45 per acre

(b) Permanent pump unit with underground main covering 56 acres in 10 days

	£
Diesel pump unit	1,005
Suction/delivery	350
1,660 yards 5 in underground main	2,400
Hydrant assemblies	115
1,350 ft of 4 in. portable main	900
1,190 ft of 3 in. sprinkler line	750
Installation cost	1,590
	7,110 or £127 per acre

For small horticultural schemes, costs can again vary from £50 per acre to over £400 per acre; the lower costs would include similar systems as used with normal agricultural schemes. The higher costs result from schemes based on a grid system of irrigation where the whole area is covered with a gridwork of small bore aluminium pipes (thus eliminating pipe moving).

(c) Grid line system with underground mains covering 14 acres

	£
Tractor pump	320
Underground main inlet	75
430 yd 4 in. underground main	355
Hydrants	150
9,300 ft of 1¼ in. grid line and equipment	1,300
Installation	300
	2,500 or £178 per acre

Schemes which require expenditure on source works such as the provision of a borehole or reservoir will cost considerably more than those quoted above; typical examples of these costs are £1,750 to £2,500 for a borehole supplying 10,000 gal an hr, and £800 to £1,200 per million gal for reservoirs.

Application Costs

The main elements of application costs are water charges, labour and pumping.

1. WATER CHARGES

The charging scheme under the Water Resources Act 1963 came into force on 1 April 1969 and the cost will vary according to quality of water and the River Authority concerned; for good quality water typical charges for surface water are 0·5p to 0·8p per 1,000 gal and for ground water 0·3p to 0·4p. Thus with water at 0·5p and 0·8p per 1,000 gal the respective costs per acre in. are 11p+18p.

2. LABOUR

The labour requirement for irrigation is an increasing problem on many farms. It is not that the labour need is particularly high but that the demand is unpredictable, being dependent on the weather, and may come at a peak time for other tasks. Average man-hours per acre in. for agricultural systems would be 2·3 and 0·75 for a typical horticultural system.

3. PUMPING

This cost will depend on the price of fuel and the pump output; with fuel at 12·6p per gal typical costs for the example schemes mentioned above would vary between 34·8p per acre in. (pump output 177 g.p.m.) and 51·9p per acre in. (pump output 84 g.p.m.).

Operating Costs

Total operating costs are the sum of the fixed annual charges for depreciation and interest based on the capital cost plus the variable annual application costs. These costs at different levels of capital expenditure can be expressed as follows:

APPENDIX V

	Agricultural system		Horticultural system	
	£	£	£	£
Capital cost per acre	50	100	150	250
Depreciation over 10 years and interest at 12 per cent	9·26	18·51	27·77	46·28
Pumping costs	1·30	1·11	1·04	1·56
Water charge	·53	·53	·53	·53
Total per acre	11·09	20·15	29·34	48·37
Per acre in. (assuming 3 in. per year)	3·70	6·72	9·78	16·12
Cost of labour (if applicable)	1·32	1·32	1·32	·52
Total (including labour but excluding maintenance) per acre in.	5·02	8·04	11·10	16·64

Appendix VI

IRRIGATION TECHNICAL DATA

One inch = 25·4 millimetre
One foot = 0·3048 metre
One acre = 4,047 square metre or 0·4047 hectare

One Imperial gallon
- = 0·16 cu. ft
- = 10 lb at 62°F
- = 1·2 U.S. gal
- = 0·004546 cubic metre
- = 4·546 litres

One cubic ft of water
- = 6·25 Imperial gal
- = 62·5 lb at 62°F
- = 0·028 cubic metre
- = 28·32 litres
- = 28·32 kg

One Cusec (cubic ft/second)
- = 375 gal per min
- = 0·028 cubic metre per second
- = 28·32 litres/sec

One inch of water
- = 22,610 gal/acre
- = 25·4 litres/square metre
- = 4½ gal/sq. yd (approx)
- = 376·8 gal/min applied to 1 acre in 1 hr

One acre-inch
- = 22,610 gal
- = 102,800 litres
- = 102·8 cubic metre

One pound per square inch pressure
- = 2·31 ft head
- = 6,895 newton/sq. metre

One horsepower = 745·7 watt.

Formulae

NOTE. THESE MUST NOT BE USED WITH METRIC UNITS.

1. Gallons per Minute Required (G.P.M.)

$$\frac{22{,}610 \times I \times A}{H \times D \times 60} = \text{gal/min}$$

I = depth in in. to be applied
A = area in acres
H = hours of operation per day
D = days required to cover area

It is usual to assume that a supply of 1 in. every 10 days is necessary over the area to be irrigated at any one period of the year. Peak demand can be met by increasing the number of hours of operation during the day.

APPENDIX VI

2. Application Rate

$$\frac{\text{G.P.M.} \times 115.5}{\text{Distance between nozzles (ft)} \times \text{Distance between distribution lines (ft)}} = \text{Application rate in./hr}$$

G.P.M. = discharge in gpm from sprayline nozzle, a single sprinkler or rain-gun

3. Water Horse-power Required (W.H.P.)

$$\frac{\text{G.P.M.} \times H}{3,300} = \text{W.H.P.}$$

G.P.M. = total gpm
H = total head in ft

4. Actual Horse-power Required (B.H.P.)

W.H.P. × Efficiency factor = B.H.P.
Efficiency factor varies with pump unit. As a guide
B.H.P. = 2 × W.H.P. approximately.

FLOW OF WATER OVER A 90° NOTCH

Head of water above base of notch (in.)	Discharge (gal per hr)
$\frac{1}{2}$	21
1	117
$1\frac{1}{2}$	322
2	654
$2\frac{1}{4}$	870

Calculation of Rate of Flow in Water Supply

Open irrigation—1 acre
Glasshouse —$\frac{1}{2}$ acre

	Type of irrigation	Total water per day (gal)	Required rate of flow (gal/hr)	Time to complete	Required pressure at nozzle
Open irrigation	1 in. on $\frac{1}{10}$ acre per day or 1 in. on $\frac{1}{5}$ acre per day	2,300 4,600	600 1,200	Completed in 4 hr Completed in 4 hr	20 p.s.i.
Trickle irrigation	Tomatoes 7,000 plants/$\frac{1}{2}$ ac Max. 3 pt per plant Av. 2 pt per plant Nozzle rate 2 pt/hr	2,500	875 (7,000 pt) $\frac{1}{2}$ of area 3,500 nozzles	$\frac{1}{4}$ ac completed in $1\frac{1}{2}$ hr $\frac{1}{2}$ ac completed in 3 hr	6 ft/head
Damping down	36 nozzles/$\frac{1}{2}$ ac 150 g.p.h./nozzle 600 sq. ft/nozzle 5 min per day or 360 nozzle/$\frac{1}{2}$ ac 40 g.p.h./nozzle 60 sq. ft/nozzle 5 min per day	450 1,200	450 (3 nozzles) $\frac{1}{12}$ of area 1,200 (30 nozzles) $\frac{1}{12}$ of area	Completed in 1 hr Completed in 1 hr	20 p.s.i. 20 p.s.i.
Winter flooding	4 in. water 40 g.p.h./nozzle 60 sq. ft/nozzle 360 nozzles	Total water gal 46,000	1,200 (30 nozzles)	$\frac{1}{12}$ area completed in $3\frac{1}{4}$ hr Whole completed in 38 hr	20 p.s.i.

IRRIGATION

Capacity of storage		Pump capacity	
Open irrigation	2,300 gal		0·75 h.p. 1,200 g.p.h. at 60 ft/head
Trickle	2,500 „	or —	1·0 h.p. 1,200 g.p.h. at 80 ft/head
Damping down	1,200 „	or —	1·5 h.p. 1,200 g.p.h. at 100 ft/head
Total	6,000 gal		

Night flow required $\dfrac{6,000}{12}$ = 500 g.p.h.
or less if some day flow is possible.

CALCULATION OF RATE OF FLOW IN WATER SUPPLY

Glasshouse—1 acre

	Type of irrigation	Total water per day gal	Required rate of flow gal/hr	Time to complete	Required pressure at nozzle
Trickle irrigation	Tomatoes 14,000 plants Max. 3 pt per plant Av. 2 pt per plant Nozzle rate 2 pt per hr	5,000	1,750 (14,000 pt) ½ of area 7,000 nozzles	1 ac completed in 1¼ hr 1 ac completed in 3 hr	6 ft/head
Damping down	72 nozzles 150 g.p.h./nozzle 600 sq. ft/nozzle 5 min per day or	900	900 (6 nozzles) 1/12 of area	Completed in 1 hr	20 p.s.i.
	720 nozzles 40 g.p.h./nozzle 60 sq. ft/nozzle 5 min per day	2,400	2,400 (60 nozzles) 1/12 of area	Completed in 1 hr	20 p.s.i.
Winter flooding	4 in. water 720 nozzles 40 g.p.h./nozzle 60 sq. ft/nozzle	Total water gal 92,000	2,400 60 nozzles 1/12 of area	1/12 of area completed in 3¼ hr 1 ac in 38 hr	20 p.s.i.

Storage capacity		Pump capacity	
Trickle irrigation	5,000 gal	2 h.p. 22,400 gal/hr 100 ft/head	
Damping down	2,400 „	1½ h.p. 2,400 gal/hr 70 ft/head	
	7,400 „		

12-hr night flow = 600 gal/hr
Less if some day flow is possible.

GLASSHOUSE IRRIGATION
TOMATOES UNDER GLASS

TRICKLE IRRIGATION

Water requirement
4 pt per plant per day — Very bright day
3 pt „ „ „ „ — Bright day
2 pt „ „ „ „ — Moderately bright day
1 pt „ „ „ „ — Dull

14,000 tomato plants per acre

Daily and hourly flow rates

Quantity per plant	1/10 ac glass			½ ac glass			1 ac glass		
	Quantity per day	Watering completed		Quantity per day	Watering completed		Quantity per day	Watering completed	
		7 hr	3½ hr		7 hr	3½ hr		7 hr	3½ hr
pt	gal	g.p.h.	g.p.h.	gal	g.p.h.	g.p.h.	gal	g.p.h.	g.p.h.
4	700	100	200	3,500	500	1,000	7,000	1,000	2,000
3	500	70	140	2,600	370	740	5,200	740	1,480
2	350	50	100	1,750	250	500	3,500	500	1,000

Note: Cameron normal trickle nozzles apply 2 pt water in 1 hr

APPENDIX VI

DAMPING DOWN

The five examples given below are based on commercially available nozzles:

Example 1

150 g.p.h. per nozzle at 20 p.s.i.
30 ft width × 20 ft spacing = 600 sq. ft per nozzle

1 ac glass $\dfrac{43,500}{600}$ = 72 nozzles = 10,800 gal/hr = 0·47 in. water/hr

½ ac glass $\dfrac{21,750}{600}$ = 36 nozzles = 5,400 gal/hr = 0·47 in. water/hr

$\tfrac{1}{10}$ ac glass $\dfrac{4,350}{600}$ = 7 nozzles = 1,050 gal/hr = 0·47 in. water/hr

Example 2

60 g.p.h. per nozzle at 15 p.s.i.
15 ft width × 8 ft spacing = 120 sq. ft per nozzle

1 ac glass $\dfrac{43,500}{120}$ = 360 nozzles = 21,600 g.p.h. = 0·95 in. water/hr

½ ac glass $\dfrac{21,750}{120}$ = 180 nozzles = 10,800 g.p.h. = 0·95 in. water/hr

$\tfrac{1}{10}$ ac glass $\dfrac{4,350}{120}$ = 36 nozzles = 2,160 = 0·95 in. water/hr

Example 3

110 g.p.h. per nozzle at 15 p.s.i.
10 ft width × 6 ft spacing = 60 sq. ft per nozzle

1 ac glass $\dfrac{43,500}{60}$ = 720 nozzles = 29,000 g.p.h. = 1·25 in. water/hr

½ ac glass $\dfrac{21,750}{60}$ = 360 nozzles = 14,500 g.p.h. = 1·25 in. water/hr

$\tfrac{1}{10}$ ac glass $\dfrac{4,350}{60}$ = 72 nozzles = 2,900 g.p.h. = 1·25 in. water/hr

Example 4

15 g.p.h. per nozzle at 15 p.s.i.
7·5 ft width × 3·5 ft spacing = 26 sq. ft

1 ac glass $\dfrac{43,500}{26}$ = 1,680 nozzles = 25,000 g.p.h. = 1·1 in. water/hr

½ ac glass $\dfrac{21,750}{26}$ = 840 nozzles = 12,500 g.p.h. = 1·1 in. water/hr

$\tfrac{1}{10}$ ac glass $\dfrac{4,350}{26}$ = 168 nozzles = 2,500 g.p.h. = 1·1 in. water/hr

Example 5

18 g.p.h. at 10 p.s.i.
40 g.p.h. at 20 p.s.i.
50 g.p.h. at 40 p.s.i.
approx 10 feet width 6 feet spacing = 60 sq. ft

1 ac glass 43,500 = 720 nozzles = 13,000 g.p.h. at 10 p.s.i. = 0·6 in. water/hr
                                 = 29,000 g.p.h. at 20 p.s.i. = 1·25 in. water/hr
                                 = 36,000 g.p.h. at 40 p.s.i. = 1·6 in. water/hr

½ ac glass $\dfrac{21,750}{60}$ = 360 nozzles = 6,500 g.p.h. at 10 p.s.i. = 0·6 in. water/hr
                                 = 14,500 g.p.h. at 20 p.s.i. = 1·25 in. water/hr
                                 = 18,000 g.p.h. at 40 p.s.i. = 1·6 in. water/hr

$\tfrac{1}{10}$ ac glass $\dfrac{4,350}{60}$ = 72 nozzles = 1,300 g.p.h. at 10 p.s.i. = 0·6 in. water/hr
                                 = 2,900 g.p.h. at 20 p.s.i. = 1·25 in. water/hr
                                 = 3,600 g.p.h. at 40 p.s.i. = 1·6 in. water/hr

WINTER FLOODING

Total water required about 4 in. = 4 × 23,000 gal per acre:
  (a) Soil infiltration rate ¼ in./hr = 5,750 gal/hr per ac = 1 in. of water applied in 4 hr
  (b) Soil infiltration rate ½ in./hr = 11,500 gal/hr per ac = 1 in. of water applied in 2 hr
  (c) Soil infiltration rate 1 in./hr = 23,000 gal/hr per ac = 1 in. of water applied in 1 hr

Note: Where the higher rates of application are used it will be necessary to divide the area into blocks so that the rates of flow do not exceed the capacity of the supply pipe or booster pump.

OUTSIDE IRRIGATION

Normal irrigation cycle

1 in. of water applied to irrigated area in 10 days equivalent to applying 1 in. of water to $\frac{1}{10}$ area per day. Assuming area is 1 acre, water applied to $\frac{1}{10}$ acre = 2,300 gal/day.

If soil infiltration rate is $\frac{1}{4}$ in. water per hr, equals $\dfrac{2{,}300}{4}$ = 600 g.p.h. for 4 hr per day.

If required irrigation cycle is

1 in. of water applied to irrigated area in 5 days equivalent to applying 1 in. water to $\frac{1}{5}$ area per day. Assuming area is 1 acre, water applied to $\frac{1}{5}$ ac = 4,600 gal/day.

If soil infiltration rate is $\frac{1}{4}$ in. water per hr, equal $\dfrac{4{,}600}{4}$ = 1,200 g.p.h. for 1$\frac{1}{4}$ hr per day, or the area could be irrigated at 600 g.p.h. for 8 hr per day, i.e. by irrigating $\frac{1}{10}$ ac in the morning and a second $\frac{1}{10}$ ac in afternoon.

OPEN-TOPPED WATER STORAGE TANKS

Circular and rectangular water storage tanks of various sizes and capacity are commercially available, for storing water for irrigation purposes. If the supply is from the public main the installation and connection must conform to the water undertakers byelaws.

Tanks are normally manufactured in corrugated galvanized steel sheets bolted together, and are provided with a waterproof lining with a minimum thickness of 0·03 in. (0·75 mm). In certain cases safety fences must be installed. Covers may be provided in order to maintain water quality. They are also obtainable in steel plate and other materials, and are normally erected on a 4 in. or 6 in. thick concrete base, suitably reinforced where required.

The following data give a selection of tanks that are available, but there are many other sizes and capacities to suit specific requirements.

Circular tanks				
Diameter		Height		Capacity
ft	in.	ft	in.	gal
10	0	4	0	2,000
12	0	5	0	3,500
16	6	6	0	7,500
18	0	5	0	8,000
20	0	10	0	19,000
24	0	7	6	21,000
36	0	7	6	49,000

Rectangular tanks						
Size				Height		Capacity
ft	in.	ft	in.	ft	in.	gal
12	0	× 12	0	4	0	3,300
18	0	× 18	0	4	0	7,500
24	0	× 18	0	4	0	10,000
24	0	× 25	0	4	0	13,400
30	0	× 30	0	4	0	21,000

(May be obtained with walls 3 ft 0 in., 4 ft 0 in. or 6 ft 0 in. in height).

Index

Advance irrigation, 97
Air-lift pumps, 57
Animal diseases, 38
Anti-transpirants, 9
Apparent specific gravity, 5
Apples, 37, 100
Application:
    costs, 130
    glasshouses, 60
    rate, 8, 54
Aquifers, 32
Asparagus, 77
Available water capacity, 3, 6
Averages:
    potential transpiration, 104 et seq.
    rainfall, 104 et seq.

Balance, water, 11
Beans, 77
Beetroot, 78
Bench:
    irrigated, 70
    warming, 73
Bitter pit, 86
Black currants, 87, 100
Blockages, 59, 68
Boreholes, costs, 130
Brassicae, 9, 74, 78–80
Brussels sprouts, 78
Bulbs, 85
Bush fruits, 87, 100

Cabbage, 74, 78
Calculations, water supply, 133
Cane fruits, 87
Capacity, field, 2, 5, 7, 60
Capillarity, 2, 6, 72
Capillary watering, 69 et seq.
Capital costs, 129
Carnations: 66, 67, 69
    water figure, 93
Carrots, 79
Cauliflowers, 74, 79
Celery, 74, 80
Centrifugal pumps, 55
Cereals, 74, 80
Charges, water, 130
Chrysanthemums; 84
    water figure, 83
Climate and water needs, 19 et seq.
Cloches: 94
    frost sprinkling 100
Clovers, 37
Cocksfoot, 37
Control valves 59
Corms, 85

Costs:
    application, 130
    boreholes, 130
    capital, 129
    equipment, 40
    grid lines, 130
    horticultural schemes, 129
    installation, 129
    mains, 129
    operating, 130
    pumping, 130
    reservoirs, 130
    solid-set systems, 101
    sprinkler lines, 129
    water application, 130
County values, water needs, 15 et seq.
Couplings, 42, 47, 52
Crop type, equipment, 40
Cropping, protected, 94 et seq.
Crops:
    growth, 74 et seq.
    moisture sensitive stages, 9
    response to irrigation, 18
Cucumbers, 66
Cultivation, soil moisture loss, 10
Cultural practices, 73

Damping down, 133–5
Damping, overhead, 64
Data:
    planning, 124–5
    technical, 132
Deficits:
    planned, 74
    soil moisture, 12
Depth, rooting, 74
Dew, 1
Diffuser nozzles, 65, 96
Dilutors: 62
    displacement, 63
    venturi, 62
Diseases, hop, 89
Distribution:
    equipment, 40
    fertilizers, 73
    pesticides, 73
Drainage: 2, 6
    impeded, 74
    valves, 59
Drip watering, 45, 68
Drop size, 8, 45
Dwarf beans, 77

Earliness, 75
Earth reservoirs, 35
Ejector pumps, 57

INDEX

Electronic leaf, 63, 72
Equalizers, pressure, 52
Equipment:
    costs, 40
    crop type, 40
    distribution, 40
    field trickle, 45
    frost damage, 59
    frost protection, 101
    glasshouses, 60 *et seq.*
    labour, 42
    lay-flat tubes, 69
    outdoor, 40
    power, 41
    soil texture, 50
    topography, 41
    water supply, 42
Estimation of irrigation needs, 111 *et seq.*
Evaporation: 2
    soil, 10
Evaporimeters, 60
Excess rainfall: 11
    map, 49
    winter, 38
Excessive watering, 75

Farm crops, 77 *et seq.*
Feeds, liquid, 91
Fertilizers, 62, 63, 73, 102
Field capacity: 2, 5, 7, 60
    determination, 5
Filters: 73
    blockage, 59
    nozzle, 66
Flooding, winter, 64, 69, 133–5
Flow calculations: 133–4
    gauging, 30
Flowers, 43, 66, 67, 69, 84–5
Formulae, 132
Frames, 94
French beans, 77
Friction losses, 46, 47
Frost:
    damage to equipment, 59
    protection, 73, 97 *et seq.*
    radiation, 98
    thermometers, 100
    warnings, 100
Fruit crops:
    85 *et seq.*
    blossom, 98
    colour, 101
    frost, 98
    soil, 86, 89
    trees, 85
Fungicides, 102

Gauge, pressure, 54, 59
Gladioli, 84
Glasshouses: 30, 60 *et seq.*, 91
    equipment, 64 *et seq.*
    frequency of watering, 62

Glasshouses:—*continued*
    irrigation, 66, 133
    moisture regime, 91
    water losses, 60, 92
    water meters, 62
    water needs, 60
    water quality, 73
Grants, government, 101
Grass: 13, 35, 45, 74, 90
    intensive irrigation, 90
    limited irrigation, 91
    nitrogen, 90
    water quality, 38
Grid lines, costs, 130
Ground water: 32
    yield, 33
Growth:
    plant, 73
    substances, 102

Head, pressure, 46, 62, 67
Herbicides, 102
Hops: 88
    diseases, 89
    response periods, 89
    soil, 89
Horticultural schemes, costs, 129
Hosepipes, 64
Hydrants, 47

Impeded drainage, 74
Infiltration, 8
Injectors, fertilizer, 63
Insecticides, 102
Installation costs, 129 *et seq.*
Instruments:
    electronic leaf, 63, 72
    evaporimeters, 60
    thermometers, 100
    lysimeters, 92
    neutron probes, 5
    photometers, 60
    tensiometers, 4, 5, 61
    tensiostats, 63
Intensive irrigation, grass, 90
Irrigated bench: 70
    plant containers, 72
    watering, 71
Irrigation:
    advance, 97
    costs, 129
    multi-purpose, 102
    objects, 8, 73
    planning, 18
    systems, glasshouses, 66

Labour requirements, 130
Lagging, 59
Latent heat, water, 98
Laterals: 48
    lay-out, 49
    movement, 50

INDEX

Lay-flat tube system, 69
Leaching: 2
    factor, 39
Leaf nutrients, 102
Leeks, 81
Lettuce, 37, 64, 81, 92, 95
Licences, 29
Lights, 94
Limited irrigation:
    fruit, 86
    grass, 91
Liquid feeds, 91
Livestock, 90
Low-level spraylines: 66
    sprinklers, 68
Lubrication, pumps, 59
Lysimeters, 92

Main pipes, 46
Mains, costs, 129
Management, 58
Marrows, 81
Maturity, 75
Maximum water needs, 19, 111 *et seq.*
Meteorological data:
    regional, 103 *et seq.*
    aspects, 10
Meters, water, 62
Mist propagation, 72
Mist sprays, 65
Moisture:
    release curves, 6
    regime, glasshouses, 91
    tomatoes, 92
    sensitive stages, 9, 75
    stress, 1
Mulching, 10
Multi-purpose irrigation, 102
Mushrooms, 96

Neutron probe, 5
Nitrogen, grass, 90
Nomenclature, 75
Non-standard plans, 126
Nozzles: 45, 52, 65
    blockage, 66, 67, 73
    diffuser, 65, 96
    filters, 66
    flow, 52
    mist, 73
    rotary, 66
    size, 53
    slotted, 65
    water pressure, 65, 133-4
Nursery stock, 84
Nutrients:
    leaf, 102
    leaching, 39

Objects of irrigation, 8
Onions, 81, 82

Operating:
    costs, 130
    pressures, 53
Orchards, 45
Oscillatory spraylines, 42
Outside irrigation, waterflow, 136
Overhead damping: 64
    spraylines, 64

Parsnips, 82
Pears, 37
Peas, 9, 37, 82
Pesticides, distribution, 73
Perennials, herbaceous, 85
Permanent wilting, 3, 5
Photometers, 60
Photosynthesis, 1
Pipes: 45 *et seq.*
    buried, 46
    friction losses, 47
    lateral, 48
    lay-out, 49, 59
    main, 46
    movement, 50
    ring main, 47
    sizes, 46, 47
    small-bore, 51
    solid-set, 51
Piston pumps, 57
Planned deficits, 74
Planning:
    data, 124
    irrigation, 18 *et seq.*
Plans, non-standard: 24, 126
    typical, 12
Plant containers, bench, 72
Plants:
    growth, 73
    requirements, 12
    response, 74
Plums, 37
Pollution, water, 37
Polythene tunnels, frost, 100
Pot plants, 66
Potatoes: 35, 37, 82, 83
    early, 35, 82
Potential transpiration: 11, 12
    averages, 104 *et seq.*
Power supply:
    pumps, 58
    units, 58
Precipitation patterns, 53
Pressure:
    drop, 46
    equalizers, 52
    gauge, 54, 59
    head, 46, 62, 67
    operating, 53
Pre-wetting, frost, 97
Priming pumps, 55, 59
Probe, neutron, 5

INDEX

Protected cropping: 94
    frost protection, 100
    trickle irrigation, 96
Pumping:
    costs, 130
    duty, 55
Pumps: 55, 63
    air leaks, 58
    air-lift, 57
    capacity, 134
    centrifugal, 55
    costs, 129
    ejector, 57
    lubrication, 59
    operation, 59
    piston, 57
    power supply, 58
    priming, 55, 59
    siting, 57
    starting, 59
    strainers, 57
    submersible, 57
Pulses, 74

Radiation:
    frosts, 98
    solar, 1, 60, 92
Radish, 83
Rainfall: 11
    county averages, 104 *et seq.*
    distribution, 103
    excess, 11, 38
    percentages, 103
Rain-guns, 45
Raspberries: 87
    yield, 88
Red currants, 37
Regional meteorological data, 104 *et seq.*
Reservoirs: 29
    costs, 130
    earth, 35
    impounding, 35
    Safety Provisions Act 1930, 35
    seepage, 36
Response periods:
    fruit, 89
    plant, 74
Rhubarb, 83, 96
Ring mains, 47
River authority: 32
    fees, 29
Root crops: 9
    distribution, 92
    suction, 4
Rooting depth, 74
Roses, 66
Rotary nozzles, 66
Runner beans, 77
Run-off, 2

Salad crops, 42

Seasonal water needs, 13 *et seq.*
Seepage reservoir, 36
Sequence controllers, 63
Sewage, 36
Siting pumps, 57
Slotted nozzles, 65
Sodium chloride, 37
Soft fruit, 37, 88
Soil:
    1 *et seq.*, 19
    available water, 3, 6, 74
    electrical properties, 5
    evaporation, 10
    irrigation damage, 8
    moisture content, 2, 5
    moisture deficits, 12, 74
    moisture determination, 5
    moisture losses, 1, 2, 10
    moisture properties, 3, 5
    moisture release curves, 4
    moisture tension, 2, 4
    movement of water, 2, 6, 8, 67
    pores spaces, 1
    rate of irrigation, 41
    reservoir, 1, 8
    saturation, 1
    specific gravity, 5
    structure, damage to, 64
    thermal properties, 5
    water application, 41
Solar radiation, 1, 60, 92
Solarimeters, 93
Solenoid valves, 64
Solid-set systems: 45, 102
    cost, 101
Spinach, 83
Spraylines: 42
    coupling, 42
    low-level, 66
    oscillatory, 42
    overhead, 64
    support, 66
Sprays, mist, 65
Sprinkler line, costs, 129
Sprinklers: 43, 94
    drop size, 45
    frost, 98
    heads, 59
    large, 45
    lay-out, 49
    low-level, 68
    maximum numbers, 44
    orchards, 44
    overlapping, 54
    precipitation rates, 43
    water distribution, 44
Sprinkling, frost protection, 98
Standpipes, 47
Stick beans, 42
Stocking rate, 90
Stomata, 1
Stop-ends, 48

INDEX

Storage capacity: 134
    open-topped tanks, 136
    tanks, 62
    water, 35
Strainers, pump, 57
Strawberries, 37, 88
Structures, 91
Submersible pumps, 57
Sugar beet, 35, 45, 83
Surface water, 40
Swedes, 84
System management, 58

Take-off valves, 48
Tanks, storage, 62, 136
Technical data, 132
Temporary wilting, 3, 9
Tensiometers, 4, 5, 61
Tension, soil moisture, 2, 4
Tensiostats, 63
Thermometers, 100
Timing of irrigation, 9
Tomatoes: 62, 66, 69, 91, 134
    moisture regime, 92
    root distribution, 92
    water factor, 92
    water figure, 93
Topography and equipment, 41
Total water needs, 24 *et seq.*
Transpiration: 1, 2, 11, 60, 72
    altitude, 11
    coastal, 11
    county averages, 104 *et seq.*
    height corrections, 104
    meteorological factors, 9, 10, 11
    potential, 11, 12, 104 *et seq.*
    reduction, 9
Tree fruit: 14, 37, 74, 86, 100
    frost protection, 100
Trickle irrigation, 62, 66, 96, 133, 134
Tubers, 85
Tunnels, 94
Turnips, 84

Underground water, 29

Valves: 47
    automatic, 50, 52
    control, 59
    drainage, 59
    solenoid, 64
    take-off, 48
Vegetables, 9, 74, 77 *et seq.*
Venturi dilutors, 62
V-notch weir, 31

Warming, bench 73
Warnings, frost, 100

Water:
    application, 54, 60
    application bench, 71
    application soils, 41
    balance, 11
    balance, grass, 12
    biological quality, 37
    capillary movement, 2
    charges, 130
    chemical quality, 36
    control, bench, 71
    factor, tomatoes, 92
    figure, 93
    flow through pipes, 45
    ground, 33
    latent heat, 98
    meters, 62
    needs, 10 *et seq.*
    ,, analyses, 12
    ,, climate, 19 *et seq.*
    ,, county values, 15–17
    ,, estimation, 111 *et seq.*
    ,, frequency, 18, 111
    ,, glasshouses, 60
    ,, limited periods, 14, 24
    ,, maximum, 19, 111
    ,, seasonal, 13 *et seq.*
    ,, total, 24 *et seq.*

Water:
    pollution, 36
    pressure, nozzles, 65
    public supply, 35
    quality, 36
    quality, glasshouses, 73
    resources act (1963), 29, 37, 130
    spread, 67
    storage, 35
    supply, sources, 28
    surface, 30
    underground, 29
    vapour, 1
    yield of sources, 38
Waterflow, measured head, 32
Watering:
    capillary, 69
    drip, 45, 68
    excessive, 75
    rods, 97
    timing, 9
Weighing machines, 92
Weir, V-notch, 31
Wilting:
    permanent, 3
    point, determination, 5
    temporary, 3
Wind, 54
Winter flooding, 64, 69, 133 *et seq.*

HER MAJESTY'S STATIONERY OFFICE

Government Bookshops

49 High Holborn, London WC1V 6HB
13a Castle Street, Edinburgh EH2 3AR
41 The Hayes, Cardiff CF1 1JW
Brazennose Street, Manchester M60 8AS
Southey House, Wine Street, Bristol BS1 2BQ
258 Broad Street, Birmingham B1 2HE
80 Chichester Street, Belfast BT1 4JY

Government publications are also available through booksellers

ISBN 0 11 241438 9